Current Topics in Microbiology and Immunology

Volume 403

AF807275

More information about this series at http://www.springer.com/series/82

Shigekazu Nagata · Hiroyasu Nakano
Editors

Apoptotic and Non-apoptotic Cell Death

Responsible Series Editor: Tasuku Honjo

Springer

Editors
Shigekazu Nagata
Biochemistry & Immunology
WPI Immunology Frontier Research Center,
Osaka University
Osaka
Japan

Hiroyasu Nakano
Department of Biochemistry
Toho University School of Medicine
Tokyo
Japan

ISSN 0070-217X ISSN 2196-9965 (electronic)
Current Topics in Microbiology and Immunology
ISBN 978-3-319-79553-9 ISBN 978-3-319-23913-2 (eBook)
DOI 10.1007/978-3-319-23913-2

Printed on acid-free paper

This Springer imprint is published by Springer Nature
The registered company is Springer International Publishing AG
The registered company address is: Gewerbestrasse 11, 6330 Cham, Switzerland

Preface

Apoptosis is a prototype of regulated cell death and is executed by sequential activation of proteases called caspases. Once cells undergo apoptosis, dead cells are rapidly engulfed by phagocytes to prevent the release of noxious materials from dying cells. Apoptotic cells or phagocytes engulfing apoptotic cells release various factors that inhibit inflammation, stimulate cell proliferation, or induce the immune tolerance. Thus, apoptosis is believed to play a crucial role in the animal development and tissue homeostasis. Recently, various other types of cell death or non-apoptotic forms of cell death have been identified. They are called necroptosis, pyroptosis, and ferroptosis. These non-apoptotic death processes seem to play an important role in ischemic reperfusion injury, inflammation, and elimination of virus infection. In this book, the current advance of cell death research is discussed.

Osaka, Japan — Shigekazu Nagata
Tokyo, Japan — Hiroyasu Nakano

Contents

Prix Fixe: Efferocytosis as a Four-Course Meal

Jennifer Martinez

Abstract During development, stress, infection, or normal homeostasis, billions of cells die on a daily basis, and the responsibility of clearing these cellular corpses lies with the phagocytes of innate immune system. This process, termed efferocytosis, is critical for the prevention of inflammation and autoimmunity, as well as modulation of the adaptive immune response. Defective clearance of dead cells is characteristic of many human autoimmune or autoinflammatory disorders, such as systemic lupus erythematosus (SLE), atherosclerosis, and diabetes. The mechanisms that phagocytes employ to sense, engulf, and process dead cells for an appropriate immune response have been an area of great interest. However, insight into novel mechanisms of programmed cell death, such as necroptosis, has shed light on the fact that while the diner (or phagocyte) is important, the meal itself (the type of dead cell) can play a crucial role in shaping the pursuant immune response.

Contents

J. Martinez (✉)
Immunity, Inflammation, and Disease Laboratory, National Institute
of Environmental Health Sciences, 111 T.W. Alexander Drive, Research Triangle Park,
NC 27709, USA
e-mail: jennifer.martinez3@nih.gov

Current Topics in Microbiology and Immunology (2017) 403:1–36
DOI 10.1007/82_2015_467

Published Online: 18 August 2015

1 Introduction

The phagocytic cells of our innate immune system act as surveyors of the environment, constantly patrolling the body for unwanted, unneeded, and unexpected components and ridding them in a timely and orderly fashion. The ancient, evolutionarily conserved pathway of phagocytosis ("the cellular process of eating") has been at the vanguard of immunology, developmental biology, and cellular biology since its nineteenth-century discovery (and 1908 Nobel Prize in Physiology and Medicine) by Ilya Metchinkoff and Paul Ehlirch (Krysko and Vandenabeele 2010). While clearance of invading pathogens is indeed a necessary function of phagocytes, the sensing, recognition, and removal of cellular corpses are a critical role that phagocytes play during times of development, cellular homeostasis, and stress (Nagata et al. 2010).

The formation of a "wild-type," functioning organism is, in actuality, a process wrought with waste. A multitude of extra cells are generated during development, only to unceremoniously undergo programmed cell death (described below) and be cleared by phagocytes (Green 2011). During the development of *Caenorhabditis elegans*, a total of 1090 cells are generated, and 131 of them are destined for death (Kinchen 2010). Indeed, this cell death is critical for the correct development of the organisms, as animals deficient for a variety of caspases, endoproteases that mediate apoptotic cell death, are grossly malformed and often embryonic lethal (McIlwain et al. 2015). While the generation and subsequent destruction of these cells are necessary for proper development, as well as normal cellular homeostasis, wound healing, and immune responses in the adult organism, the ruin left in its wake would be catastrophic if not for the efficient work of the phagocytic system (Savill et al. 2002; Peter et al. 2010).

Despite the constant turnover of cells through programmed cell death mechanisms (not to mention those induced to die via stress or infection), it is rare to observe apoptotic cells under normal physiological conditions. Considering the average one million adult human cells that undergo apoptosis every second, one must truly appreciate the magnitude of the job facing phagocytes (Ravichandran 2010). Moreover, as a reoccurring and normal event in the life span of an organism, this process of dead cell clearance must occur in a quiescent manner, so as to not inappropriately alert the immune system (Hart et al. 2008).

In this chapter, we will explore efferocytosis not only as a process of cleanup, but also a critical regulator of the immune response. While the manner and efficiency in which dead cell cargo is degraded and processed by the phagocyte are important, the type of dead cell cargo is engulfed can also play a role in how the phagocyte responds.

2 Prepping of the Meal: Types of Cell Death

Death is a part of life. The process of generating, maintaining, and protecting a multicellular organism throughout its lifetime requires the creation and destruction of billions of cells. While damage can certainly cause unwanted cellular death, most cellular death is genetically programmed, and perturbations in these genetic programs can promote cell accumulation, autoimmunity, oncogenesis, attrition, and/or degeneration. Programmed cell death, such as apoptosis, necroptosis, or pyroptosis, is an active mechanism designed to sculpt, control, and aid the body in its development and survival. Like death itself, the innate immune system has tolerant systems in place to manage these morbid, yet necessary events. Death comes in a variety of flavors, some more appetizing than others.

2.1 Apoptosis

The most widely studied form of cell death is apoptosis. Apoptosis (from the Greek meaning "falling off") is genetically programmed cellular suicide and involves the coordinated dismantling of intracellular components designed to prevent inflammation and limit damage to the surrounding environment (Green 2011). During apoptosis, the plasma membrane forms "blebs," allowing membraned fragments containing intracellular contents to separate from the larger, dying cells as "apoptotic bodies." This blebbing is a morphological characteristic of apoptosis and also a key mechanism for confining danger-associated molecular patterns (or DAMPs) and avoid alerting the immune system (Green et al. 2009). Another hallmark of apoptosis is DNA fragmentation and chromatin condensation (Green 2011). These and other characteristics of apoptosis are largely controlled by a family of cysteine proteases with endopeptidase activity called caspases (Taylor et al. 2008).

Caspases exist only in the animal kingdom and are broadly grouped into initiator caspases (caspase-8 and caspase-9), executioner caspases (caspase-3, caspase-6, and caspase-7), and inflammatory caspases (human caspase-1, human caspase-4, and human caspase-5; rodent caspase-1 and rodent caspase-11) (Taylor et al. 2008; Green 2011). The main function of caspases is to cleave proteins to ensure an efficient yet rapid cell death. How then does the cell tolerate such a lethal family of proteases? First of all, caspases only cleave at specific sequence residues that end in aspartate residues (hence the "asp" of caspase) (Kumar 2007). Secondly, caspases exist in inactive forms, requiring dimerization (initiator caspases) or cleavage (executioner caspases) to gain activity (Kumar 2007; Taylor et al. 2008). Thirdly, caspases are not the only proteins involved in regulation of apoptosis (discussed below).

Upstream of apoptotic signaling events, initiator caspases exist as inactive monomers that must be dimerized to become active. This process is known as an "induced proximity model" and results in autocatalytic cleavage and stabilization of the dimer (Muzio et al. 1998; Boatright et al. 2003). Once active, initiator caspases are capable of cleaving numerous targets, most notably the executioner caspases (Green 2011). Executioner caspases exist as inactive dimers and cleavage by proteases, mainly initiator caspases, occurs between the large and small subunits, resulting in a conformational change that brings the two active sites of the executioner caspase dimer together to create a functional mature protease. These active executioner caspases can now cleave and activate other executioner caspases, leading to a rapid feedback loop to facilitate apoptosis (Riedl and Shi 2004).

What triggers this cascade of lethal events? Depending on the adapters and initiator caspases involved, most apoptotic programs fall into either the intrinsic or the extrinsic category. The major mechanism of apoptosis in mammals is the intrinsic or mitochondrial pathway. This pathway is activated by a variety of stress-inducing stimuli, including growth factor deprivation, cytoskeletal disruption, DNA damage, accumulation of unfolded proteins, and hypoxia, as well as developmental signals that instruct cells to die, such as hormones (Brenner and Mak 2009; Green 2011). These signals converge on the mitochondria, where the balancing act between pro-apoptotic and anti-apoptotic members of the BCL2 family orchestrates mitochondrial outer membrane permeabilization (MOMP) to release cytochrome c and other deadly sequestered proteins from within the mitochondria (Tait and Green 2010). The intricacies of BCL2 family interactions are discussed in detail in a number of other sources (Tait and Green 2010; Green 2011; Llambi and Green 2011; Llambi et al. 2011).

The result of MOMP and the release of cytochrome c is the formation of the apoptosome, a multimeric complex comprised of cytochrome c, caspase-9, and the adaptor APAF-1. The formation of the apoptosome activates caspase-9, which leads to the activation of the downstream executioner caspases, such as caspase-3 and caspase-7 (Shiozaki et al. 2002). The importance of the intrinsic apoptotic pathway is highlighted by the developmental phenotypes of gene-targeted mice deficient for components of this pathway. Unsurprisingly, mice deficient for caspase-9, APAF-1, or caspase-3 suffer from large brain outgrowths, characterized by reduced neuronal cell apoptosis, and subsequent perinatal lethality (Colussi and Kumar 1999). Thus, the fate of proper vertebrate development lies in the ability of one organelle, the mitochondria, to know when to maintain its integrity and when to release the mediators of death.

The extrinsic pathway of apoptosis, however, is triggered by signals that engage extracellular death receptors (DR). Signals such as tumor necrosis factor (TNF),

CD95-ligand (CD95-L or Fas-L), and TNF-related apoptosis-inducing ligand (TRAIL) bind the DRs TNF receptor-1 (TNFR1), CD95 (or Fas), and TRAIL-R1/2 (DR4/5), respectively (Green 2011). Engagement of a DR ligand with its cognate receptor results in the recruitment of pro-caspase-8 to the death-inducing signaling complex (DISC) formed at the cytoplasmic tail of the engaged DR that also includes the adaptor proteins FADD or TRADD. Recruitment of these caspase-8 monomers results in dimerization and activation, leading to activity of caspase-3 and caspase-7 (Ashkenazi and Dixit 1998; Boatright et al. 2003). Caspase-8 activity can also cleave and activate BCL2 family proteins to trigger the intrinsic apoptotic pathway (discussed above) to induce efficient cell death (Tait and Green 2010; Green 2011).

Perhaps the most important result of apoptosis is the careful organization and packaging of potentially immunogenic cellular component into discreet, tolerogenic pieces. While it is easy to appreciate the appropriate execution of apoptosis in terms of proper development and suicide of infected cells, the real victory is sustained cellular renewal in the absence of immune activation. As described below, not all forms of cell death are tolerated as well.

2.2 *Necrosis*

While apoptosis is considered a genetically controlled and immunologically silent mechanism of cell death, necrosis has been characterized as a passive type of cell death, resulting in organelle swelling and cellular explosion that uncontrollably releases inflammatory cellular contents (Nikoletopoulou et al. 2013). Mechanistically, classical necrosis is typically not associated with a genetic program and occurs independently of caspase activation (Leist and Jaattela 2001). Unlike apoptosis, which is critical for proper development, necrosis is thought to mediate cellular death in response to catastrophic damage or pathology, including infarction, mechanical trauma, ischemia, frostbite, and animal venom (Raffray and Cohen 1997). Likewise, apoptotic cells that are not efficiently cleared by phagocytes can undergo secondary necrosis, a process that occurs completely independently of any apoptotic machinery.

Various molecules in our cells are pro-inflammatory if they are released from cells. Collectively referred to as damage-associated molecular patterns (DAMPs) or alarmins, they can activate neighboring macrophages and dendritic cells through TLR signaling and other mechanisms (Kono and Rock 2008). Morphologically, necrotic cells are characterized by cellular swelling (oncosis), nuclear distension, and plasma membrane rupture (Green 2011; Nikoletopoulou et al. 2013). This explosion of cellular contents, including proposed DAMPs such as HMGB1, HDGF, nucleotides, metabolites, and uric acid, is typically associated with an inflammatory reaction (Nikoletopoulou et al. 2013). Importantly, DAMP release and recognition might help alert the immune system to a cell-death-inducing pathogen, but if triggered inappropriately, it can also have deleterious effects, including autoimmunity. Importantly, both the extent and type of cell death represent major means of regulating DAMP release. When one considers the trauma that triggers necrotic death, it is unsurprising that inflammation follows, whereas apoptosis, occurring in a structured setting during carefully timed points in development, proceeds without alarm. Our understanding of necrotic cell death is evolving though, as a new programmed form of necrosis, termed necroptosis, has recently been described.

2.3 *Necroptosis*

Like its name suggests, necroptosis is the marriage between the programmed nature of apoptosis and the morphological features of necrosis. As mentioned previously, mice deficient for caspase-3 or caspase-9 die perinatally and are characterized by an excess of neuronal cells and large brain outgrowths (Colussi and Kumar 1999). Intriguingly, mice deficient for caspase-8, or its anti-apoptotic homologue, FLIP, die at embryonic day 10.5 and have embryonic vascular, cardiac, and hematopoietic defects (Varfolomeev et al. 1998). Additionally, pharmacological inhibition of caspase-8 or knockdown of caspase-8 or FLIP via short interfering RNA (siRNA) sensitizes fibroblasts to TNF-induced necrotic death (Vercammen et al. 1998). This surprising observation, wherein the absence of a pro-apoptotic gene resulted in a deficit of cells, led researchers to examine alternate roles for caspase-8.

In the absence of caspase-8, ligation of the death receptor pathway of apoptosis, such as the TNF-TNFR pathway, can result in necrotic cell death and requires the kinase activity of receptor-interacting protein kinase-1 (RIPK1) and RIPK3 (Oberst et al. 2011; Weinlich and Green 2014). Strikingly, the embryonic lethality of the caspase-8-deficient mouse is fully rescued by co-ablation of receptor-interacting protein kinase-3 (RIPK3), and caspase-8/RIPK3 double knockout mice are developmentally normal, but develop a severe lymphoaccumulative disorder resembling that of mice or humans lacking Fas or FasL (Bidere et al. 2006; Kaiser et al. 2011; Oberst et al. 2011).

How then does engagement of the extrinsic pathway of apoptosis result in necroptosis? RIPK1 plays a paradoxical role in the survival of the cell. Ligated death receptors, such as TNFR1 (or TLR engagement of TRIF (Feoktistova et al. 2011; Dillon et al. 2014)), promote the recruitment and deubiquitination of RIPK1 to the adaptor TRADD. Together with TRAF2 and the ubiquitin ligases cIAP1, cIAP2, and LUBAC, this complex, termed "Complex I," can activate the NF-κB signaling pathway, which can induce the expression of FLIP (Zhang et al. 2000; Micheau et al. 2001; Newton 2015). Subsequent formation of a cytoplasmic "Complex II" or "Ripoptosome" containing RIPK1, FADD, and caspase-8 drives apoptotic signaling upon the ligation of death receptors. However, in the absence of extrinsic apoptotic machinery, RIPK1 (or more specifically, the kinase activity of RIPK1) can promote necroptosis via RIPK3 activation. Therefore, while it has been known that FLIP blocks caspase-8-mediated apoptosis, the catalytically active complex of FADD, caspase-8, and FLIP also blocks signaling for necroptosis (Oberst et al. 2011; Newton 2015). Furthermore, inactive RIPK1 can block RIPK3 activity, even when RIPK3 is activated independently of RIPK1 (Oberst et al. 2011).

The activation of RIPK1 and RIPK3 is not the final executioner of the necroptosis pathway, however. Necroptosis also depends on the RIPK3-mediated phosphorylation of the pseudokinase, mixed lineage kinase-like (MLKL) (Kaiser et al. 2013; Rodriguez et al. 2015). Phosphorylated MLKL induces a conformational change that allows for its oligomerization and interaction with the plasma membrane by binding to phosphatidylinositol lipids to directly disrupt membrane integrity (Wang et al. 2014; Rodriguez et al. 2015).

The involvement of RIPK1 adds another layer of complexity. In addition to its role in necroptosis, the kinase activity of RIPK1 is required for a variety of innate immune signaling pathways, such as Toll-like receptors (TLRs), interferons, and the RIG-I-MAVS pathway (Dillon et al. 2014; Weinlich and Green 2014). Whereas ablation of RIPK3 rescued caspase-8-deficient animals, RIPK1/caspase-8 double knockout mice die perinatally at day 1, similar to the RIPK1-/- mice (Dillon et al. 2014). During embryonic development, RIPK1 can trigger TNFR-mediated necroptosis in animals lacking apoptotic machinery, such as FADD, FLIP, or caspase-8, and RIPK1-mediated lethality at later developmental stages that could be mediated by similar signals (termed "signal 1"). Postnatally, RIPK1 can serve two functions: (1) RIPK1 is required to prevent a TNFR1-induced apoptosis, possibly due to its role in NF-κB activation and subsequent FLIP upregulation; and (2) RIPK1 is also required to prevent RIPK3-dependent lethality promoted by other "signal 2," such as TRIF and IFN (Dillon et al. 2014). The balance and contribution of RIPK1 to cell survival, cell death, and immunity is certainly multifaceted and is discussed in elegant detail in a recent review (Weinlich and Green 2014; Newton 2015).

In addition to the role of RIPK1 in directly regulating inflammatory pathways, necroptosis has been reported to itself be an immunogenic type of cell death, and its inhibition, either genetically or with small-molecule inhibitors, has been demonstrated to lessen disease severity in several mouse models (Weinlich and Green 2014; Newton 2015). Furthermore, necroptosis is very common in vivo, not only in physical traumas, but also mainly in diverse forms of neurodegeneration, and death inflicted by ischemia or infection. Similar to classical necrosis, necroptosis results in the release of intracellular danger signals into the extracellular milieu, and these components can stimulate the immune system. Indeed, animals deficient for the necroptotic pathway are protected from various models of inflammation, such as chemically induced pancreatitis, intestine and skin inflammation, or ischemic reperfusion injury (Linkermann et al. 2013; Weinlich et al. 2013; Wu et al. 2013).

2.4 Pyroptosis

Previously, we discussed the caspases as mediators of apoptosis, but some caspases, such as caspase-1 and caspase-11 (or caspase-5 in humans), have roles in non-apoptotic, biological processes. The best-described function for caspase-1 is its key role in the processing of inactive IL-β and IL-18 into mature inflammatory cytokines. Additionally, excessive caspase-1 activity can cause pyroptosis, a non-apoptotic type of programmed cell death (LaRock and Cookson 2013; Jorgensen and Miao 2015). Execution of pyroptosis differs from apoptosis both at the biochemical and at the morphological level. Although caspase-1 can trigger apoptosis, caspase-1-mediated pyroptosis does not result in the cleavage of substrates of typical caspases. Rather, activated caspase-1 activates caspase-7 (Bergsbaken et al. 2009). Unlike apoptosis, MOMP is typically not associated with pyroptosis. Pyroptosis is characterized by lysis of the plasma membrane and the release of pro-inflammatory intracellular contents. Interestingly, nuclear DNA undergoes extensive fragmentation, similar to that observed in apoptosis, although the mechanism by which this occurs remains unknown (LaRock and Cookson 2013).

Regardless of its pro-inflammatory or pro-pyroptotic outcome, caspase-1 is activated by dimerization at complexes termed inflammasomes that form in the cytosol and detect a diverse repertoire of pathogenic molecules, including bacterial toxins and viral RNA (Henao-Mejia et al. 2012). However, how active caspase-1 kills a cell remains a complete mystery, in the sense that no key substrates have been identified that would account for this rapid deadly event. We can speculate, though, on the evolutionary roots of caspase-1's dual roles. Bacterial and viral pathogens can subvert caspase-1-mediated IL-1β and IL-18 processing, thus dampening the host inflammatory response and facilitating infection (Bergsbaken et al. 2009; Jorgensen and Miao 2015). Could pyroptosis that represents a strategy by the infected host redirect the activity of caspase-1 toward killing the cell? Alternatively, it is possible that pyroptosis is required for the release of the mature, inflammatory cytokines.

Similar to other types of lytic cell death discussed above, pyroptosis might indeed be pro-inflammatory, as it results in the release of intracellular danger signals. While caspase-1 activation is required for cell death in a variety of experimental settings, including in the immune system, the cardiovascular system, and the central nervous system, mice deficient for caspase-1 develop normally, implying that this protease is redundant in vivo during development (Green 2011; McIlwain et al. 2015). It was recently reported, however, that caspase-1-deficient mice generated from strain 129 embryonic stem cells also harbor a mutation in the caspase-11 locus, and so are in fact caspase-1/caspase-11 double knockout mice (Kayagaki et al. 2011). While it is clear that inflammation mediated by caspase-1 (or caspase-11) is a critical component of pathogen clearance and sepsis, what remains to be elucidated is the role of pyroptosis in vivo inflammatory pathologies, and whether a pyroptotic death is handled or recognized differently by the innate immune system.

3 The Dining Experience: The Mechanisms of Efferocytosis

The meal is prepared, the table is set, but how do the diners know when and where to go? Phagocytes are not typically within close proximity to the dying cells, so they must be recruited and enticed to their cellular meal (Peter et al. 2010). But recruitment is just the beginning. Efferocytosis is a carefully orchestrated process wherein phagocytes, both professional (macrophages and dendritic cells) and non-professional (epithelial cells), migrate toward areas of cell death, recognize and engage cellular corpses via surface molecules specific to dead cells, and internalize, usually in its entirety, a dead cell for degradation and processing (Parnaik et al. 2000; Poon et al. 2014). For apoptotic cells, this is an immunologically tolerable event, as phagocytes recognize cells that have undergone this cell death program as biologically inert and dispose of these corpses before they release their potentially immunogenic intracellular contents, such as DNA (Nagata et al. 2010).

We now recognize that dying cell clearance is an important factor in many different disease models, including SLE, atherosclerosis, and Alzheimer's disease (Camins et al. 2008; Ravichandran 2010). Thus, there has been a tremendous effort to understand the specific steps by which dying cells are recognized and cleared, as well as how these processes shape the immune response. Efferocytosis can be generally categorized into 4 steps: (1) the recruitment of phagocytes by "find-me" signals; (2) the recognition and engagement of "eat-me" signals; (3) the engulfment of the cellular corpse (Fig. 1); and (4) the processing, degradation, and immune response to the engulfed corpse (Fig. 2).

3.1 Perusing the Menu: "Find-Me" Signals

Even in tissues with high rates of cellular turnover, such as the thymus or bone marrow, uncleared apoptotic cells are rarely observed. Our earliest evidence of the phagocytic program of clearing dead cells comes from studies done in *C. elegans*, wherein phagocytes were recruited to sites of cell death and cleared dying cells before apoptosis (and hence complete death) was even fully executed (Reddien and Horvitz 2000; Hoeppner et al. 2001). Given that phagocytes do not often regularly reside in the tissues they must patrol, apoptotic cells must "advertise" their presence to phagocytes, essentially expediting their own clearance (Elliott et al. 2009).

Apoptotic cells release several distinct molecules, termed "find-me" signals, to attract phagocytic cells via a chemotactic gradient. To date, several potential "find-me" signals have been reported, although their relevance has not always been validated in vivo (Hochreiter-Hufford and Ravichandran 2013). B cells in germinal centers undergo increased rates of apoptosis during affinity maturation, and these apoptotic B cells release the membrane-associated molecule CX3CL1 (or fractalkine) in small vesicles or microparticles. This classical chemokine is sensed by

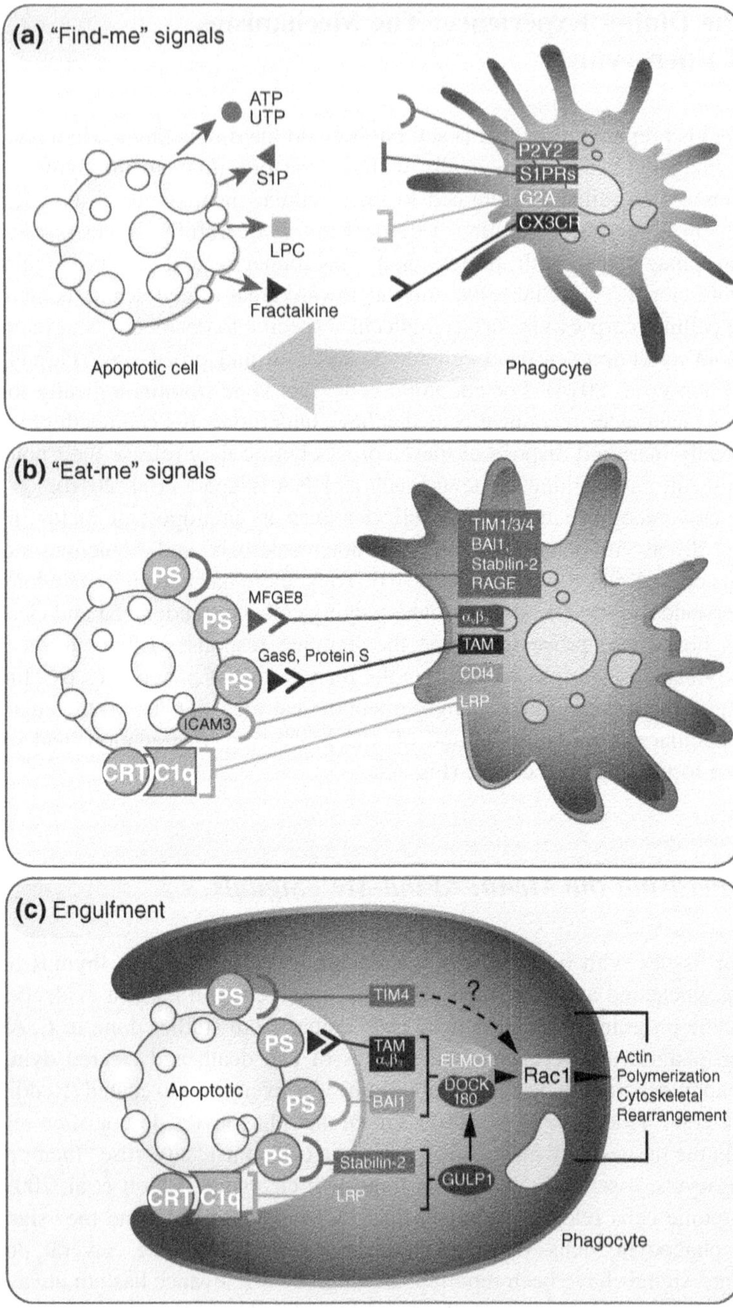
(a) "Find-me" signals
ATP
UTP
S1P
LPC
Fractalkine
P2Y2
S1PRs
G2A
CX3CR
Apoptotic cell
Phagocyte
(b) "Eat-me" signals
TIM1/3/4
BAI1,
Stabilin-2
RAGE
PS
MFGE8
Gas6, Protein S
TAM
CD14
LRP
ICAM3
CRT
C1q
(c) Engulfment
TIM4
?
TAM
ELMO1
DOCK 180
Rac1
BAI1
Apoptotic
Stabilin-2
LRP
GULP1
Actin
Polymerization
Cytoskeletal
Rearrangement
Phagocyte

◀ **Fig. 1** The recruitment of phagocytes to sites of cell death and recognition and engulfment of apoptotic cells by phagocytes. **a** Apoptotic cells (or other dying cells) release "find-me" signals, such as ATP, UTP, S1P, LPC, or fractalkine, that act to recruit phagocytes to sites of cell death. Phagocytes sense these "find-me" signals via cognate receptors (P2Y2, S1PRs, G2A, and CXCR3, respectively). **b** Phagocytes employ a system of receptors and bridging molecules to recognize and engage apoptotic cells (or other dying cells) via "eat-me" signals exposed on apoptotic cell surfaces. The most common "eat-me" signal is phosphatidylserine (or PS), which engages the PS-specific receptors, TIM1, TIM3, TIM4, BAI1, stabilin-2, and RAGE, as well as the PS-specific bridging molecules MFG-E8, Gas6, and protein S. These bridging molecules engage other surface engulfment receptors ($\alpha_v\beta_3$ or TAM) to facilitate uptake. Other "eat-me" signals, such as calreticulin (CRT) and ICAM3, exist and mediate recognition and engulfment via the receptors LRP (via C1q) and CD14, respectively. **c** Once the engulfment receptors are engaged, actin polymerization and cytoskeletal rearrangement are initiated via activation of the Rac1 pathway. While some engulfment receptors utilize the ELMO1/DOCK180 complex ($\alpha_v\beta_3$, TAM, stabilin-2, LRP), the mechanism by which TIM4 activates the Rac1 pathway is unknown. Perturbations at any step of this process can result in inflammation and autoimmunity

CX3CR1 and mediates the migration of macrophages to the dying cells. However, mice deficient for CX3CR1 do not display a defect in apoptotic cell clearance in their germinal centers, indicating that other factors function to recruit phagocytic cells (Truman et al. 2008).

Lysophosphatidylcholine (LPC) is generated and released by the caspase 3-dependent activation of phospholipase A and was the first "find-me" signal of lipid origin (Lauber et al. 2003). LPS is sensed by the G-protein-coupled receptor G2A, and this interaction can stimulate macrophage recruitment (Peter et al. 2008). Similarly, the lipid sphingosine-1-phosphate (S1P) is secreted by apoptotic cells and sensed by multiple G-protein-coupled receptors S1P-R1-5 to mediate phagocyte chemotaxis (Gude et al. 2008). These lipids, however, are present in the circulation at a concentration higher than that released by apoptotic cells, and their function in vivo has not been assessed. Therefore, their activity is likely to be merely local.

Perhaps the most promising candidates as "find-me" signals are nucleotides. ATP and UTP are released in a caspase-dependent manner via activation of pannexin 1 channels (Chekeni et al. 2010) and are detected by phagocytic cells via purinergic receptors, like P2Y2. Moreover, disruption of the nucleotide/P2Y2 interaction results in an accumulation of apoptotic thymocytes following glucocorticoid treatment in vivo (Elliott et al. 2009). It should be noted that the release of nucleotides from apoptotic cells, while an active process, is significantly smaller (less than 2 % of intracellular ATP levels) than the release that occurs during necrosis (Ravichandran 2010). Furthermore, these released nucleotides are easily degraded by extracellular nucleotidases. Thus, nucleotides most likely act to recruit tissue-resident phagocytes in a short-range capacity.

A number of caveats exist for the theory of "find-me" signals. Many cell types of different origins and function express receptors for these "find-me" signals, yet the vast majority of cells recruited to sites of cell death are macrophages. Is recruitment based on a synergistic effect of these signals or do additional signals exist to regulate migration? Lactoferrin, a glycoprotein released by apoptotic cells, has been

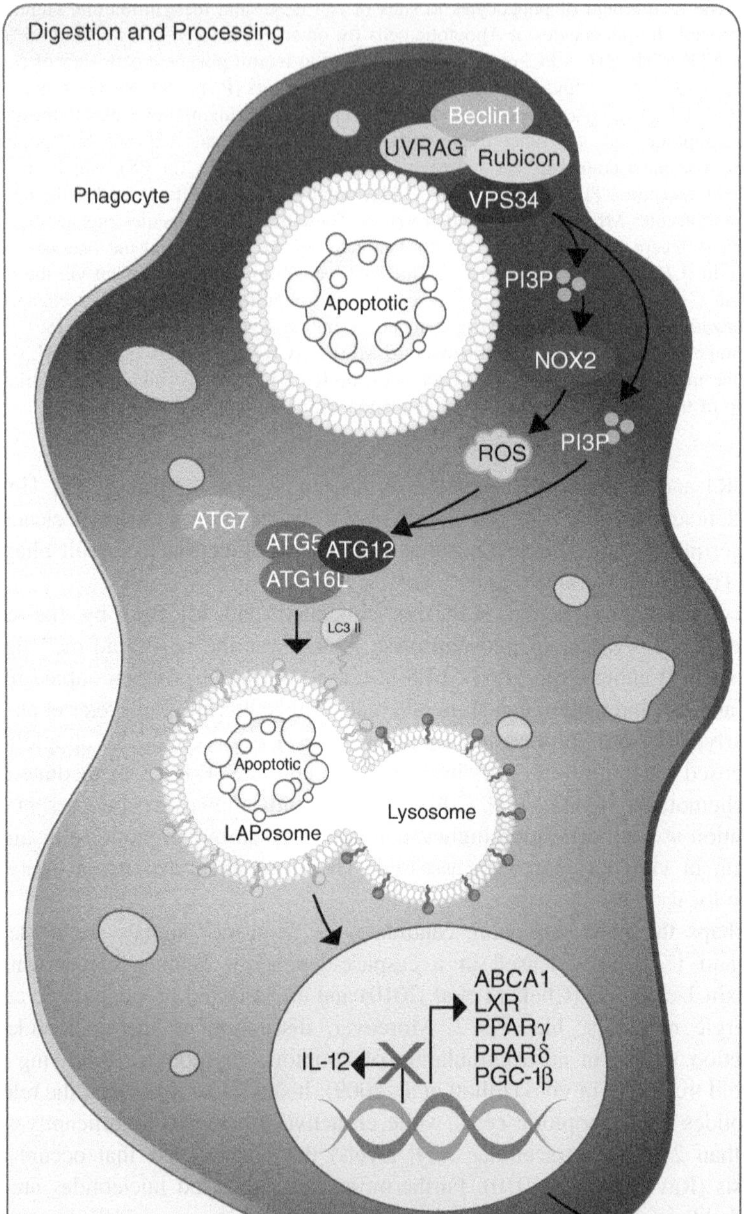
Digestion and Processing
Beclin1
UVRAG
Rubicon
VPS34
Phagocyte
Apoptotic
PI3P
NOX2
ROS
PI3P
ATG7
ATG5
ATG12
ATG16L
LC3 II
Apoptotic
LAPosome
Lysosome
ABCA1
LXR
PPARγ
PPARδ
PGC-1β
IL-12
IL-10
TGFβ

◀ **Fig. 2** The processing and digestion of engulfed apoptotic cells utilizes LC3-associated phagocytosis and promotes an anti-inflammatory response. Upon engulfment of apoptotic cells (or other dying cells), components of the LC3-associated phagocytosis (LAP) pathway are recruitment to dead cell-containing phagosome (or LAPosome). The Class III PI3 K complex, composed of Beclin-1, VPS34, UVRAG, and Rubicon, is critical to the sustained and localized production of PI(3)P at the LAPosome. PI(3)P serves two roles—the recruitment of the downstream autophagic/LAP machinery (such as ATG5, ATG12, ATG16L, ATG7) and stabilization of the NOX2 complex for the production of ROS. Of note, Rubicon is also required for the stabilization of the NOX2 complex. Both ROS and PI(3)P are required for lipidation and translocation of LC3-II to the single membrane of the LAPosome, and LC3-II is required for fusion to the lysosome and maturation of LAPosome. The anti-inflammatory effects of efferocytosis are mediated by the activity of lipid and cholesterol sensors, such as ABCA1, LXR, PPARγ, PPARδ, and PGC-1β, leading to the production of anti-inflammatory mediators, IL-10 and TGFβ. Pro-inflammatory mediators, such as IL-12, are actively repressed. Perturbations in this process can result in inflammation and autoimmunity

shown to act as a "keep-out" signal, excluding neutrophils and eosinophils from sites of cell death (Bournazou et al. 2009, 2010). However, the lactoferrin-deficient animal model has yet to be characterized, so its in vivo role remains unknown. Additionally, the relatively low concentration of these "find-me" signals begs the question—what is the range of chemotactic gradients and how are circulating phagocytes recruited to sites of cell death (Elliott et al. 2009)? While these signals are often caspase-dependent (and hence apoptosis-dependent), these molecules are also released during other forms of cell death, such as necrosis or necroptosis (Iyer et al. 2009), often in greater quantities than in apoptosis. What are the migratory and inflammatory consequences of these disparate levels of potential "find-me" signals? Indeed, some of these "find-me" signals, such as ATP, can also act as a danger-associated molecular pattern (DAMP), thereby alerting the innate immune system. Necrosis, for example, results in the release of uric acid, which acts as both a "find-me" signal and an activating factor of inflammasomes (Kono et al. 2010). Do low levels of "find-me" signals promote macrophage migration, while higher concentrations stimulate a pro-inflammatory response?

Finally, in addition to chemotaxis, what other roles do these "find-me" signals play? There is evidence that sensing of these "find-me" signals, such as ATP or S1P, activates and prime phagocytes, increasing their phagocytic capability (Hanayama et al. 2004). These and other questions remain to be answered as studies unravel the mechanisms behind the active migration of phagocytes to their meal.

3.2 *It is All in the Presentation: "Eat-Me" Signals*

The ability to distinguish self from nonself is the defining hallmark of our immune system. Likewise, the ability to distinguish living cells from dead cells is critical to development, immunity, and the prevention of unwanted inflammation. How then do phagocytes, actively recruited to sites of cell death by "find-me" signals, target dying cells, while leaving healthy cells unperturbed? It stands to reason that the process of dying transforms healthy cells into targets for engulfment, rendering them distinguishable from living cells.

Lipid bilayers comprise the core structure of the plasma and organelle membranes, and the unique makeup of these different membranes confers distinct protein folding and permeability properties (Leventis and Grinstein 2010). Lipid distribution differs not only among membranes but frequently also between the two leaflets of the bilayer. Notably, the composition of the plasma membrane is an asymmetrical distribution of lipids, wherein the lipid phosphatidylserine (PS) is actively confined to the inner leaflet in viable cells (Balasubramanian and Schroit 2003). However, during apoptosis, PS is rapidly externalized in a caspase-dependent manner. This exposure occurs not only in mammals, but also in *C. elegans* (Venegas and Zhou 2007) and Drosophila (van den Eijnde et al. 1998). The calcium-mediated cation channel TMEM16F has been shown to mediate lipid scrambling (Suzuki et al. 2010), and recent studies have demonstrated that the scramblase Xkr8 is cleaved by caspase-3 and facilitates PS exposure during apoptosis (Suzuki et al. 2013). The flippase ATP11C normally transports aminophospholipids from the extracellular to the cytoplasmic side. During apoptosis, though, ATP11C is inactivated by caspase-3 cleavage, and PS remains externally exposed (Segawa et al. 2014).

Despite its relatively minor presence in most biological membranes, PS is a lipid of great physiological importance (Leventis and Grinstein 2010). Extracellularly exposed PS is the most well-characterized "eat-me" signal and an essential factor in the recognition and clearance of apoptotic cells (Balasubramanian and Schroit 2003). Phagocytes recognize exposed PS via membrane receptors, such as T cell immunoglobulin mucin receptor 4 (TIM4), brain-specific angiogenesis inhibitor 1 (BAI1), and stabilin-2 (Park et al. 2007, 2008a; Rodriguez-Manzanet et al. 2010). Additionally, there exist bridging molecules, such as milk fat globule-EGF factor 8 (MFG-E8) and Gas6, capable of recognizing PS and being recognized by phagocytic cell surface receptors such as integrin $\alpha_v\beta_3$, $\alpha_v\beta_5$, or Tryo3-Axl-Mer (or TAM) receptors (Ishimoto et al. 2000; Hanayama et al. 2002; Zizzo et al. 2012). Engagement of these receptors can result in cytoskeletal rearrangements that facilitate the engulfment of the cellular corpse (discussed below).

While a hallmark of cell death, PS is found extracellularly in low levels on living or activated cells, yet these cells are not engulfed (van den Eijnde et al. 2001). Even forced extracellular levels of PS on viable cells, via constitutively active TMEM16F, do not result in engulfment (Segawa et al. 2011). How then does a phagocyte distinguish a PS-positive dead cell, primed for clearance, from a PS-positive cell that should live to see another day? One answer may lie in the presence of "don't eat-me" signals, such as CD31, CD47, and CD61. Engagement of these molecules, expressed on viable cells, can negatively regulate phagocytosis, thus signaling to the phagocyte that this cell, while PS-positive, is not intended for clearance (Oldenborg et al. 2000; Elward et al. 2005; Poon et al. 2014).

Further, PS is not the only "eat-me" signal identified. ICAM3, oxidized LDL-like molecules, glycosylated surface proteins, and C1q bound serum proteins have all been described to act as "eat-me" signals (Ravichandran 2010; Poon et al. 2014). The translocation of calreticulin (CRT) from the endoplasmic reticulum to the plasma membrane can also serve as an "eat-me" signal and stimulate engulfment by phagocytes (Gardai et al. 2005). While efferocytosis may be regulated by the balance of "eat-me" and "don't eat-me" signals or the synergistic effect of multiple "eat-me" signals, it is clear that dead cells actively promote their own clearance to phagocytes that have evolved to recognize and remove such cells from circulation.

3.3 *Savoring the Meal: Phagocytosis of Cellular Corpses*

Efferocytosis is an intricately choreographed process requiring action by both the dying cells and the phagocyte. While the dying cell actively recruits phagocytes to sites of cell death via "find-me" signals and advertises its desire to be cleared via "eat-me" signals, the phagocyte facilitates the actual engulfment via engagement of receptors that specifically recognize these signals. As PS is the most characterized "eat-me" signal, PS receptors are the most characterized mechanism for dead cell recognition. While initial thinking hypothesized that a single PS receptor existed to mediate this recognition and phagocytosis, we now know that multiple PS receptors exist (Bratton and Henson 2008; Nagata et al. 2010; Poon et al. 2014). PS can be recognized via bona fide membrane receptors, such as Stablin-2 (Park et al. 2008a), BAI-1 (Park et al. 2007), RAGE (He et al. 2011), and TIM4 (as well as family members TIM1 and TIM3) (Miyanishi et al. 2007; Freeman et al. 2010; Rodriguez-Manzanet et al. 2010). Additionally, bridging molecules, such as MFG-E8 (Hanayama et al. 2004; Hu et al. 2009), protein S, and Gas6 (Rothlin et al. 2007; Lemke and Rothlin 2008), have been demonstrated to simultaneously recognize PS on dead cells and promote engulfment via engagement of their cognate membrane receptors. In order to facilitate phagocytosis, MFG-E8 associates with the integrins $\alpha_v\beta_3$ or $\alpha_v\beta_5$, while the Tyro3-Axl-Mer (TAM) family of receptors recognize protein S and Gas6 (Scott et al. 2001; Lemke and Rothlin 2008).

Just as other eat-me signals exist, so do their corresponding recognition receptors. For example, lectins can recognize modified glycoproteins and lipids (Ezekowitz et al. 1990); CD36, with $\alpha_v\beta_3$, can bind thrombospondin (Fadok et al. 1998b); scavenger receptors like SR-A can bind oxidized LDL-like moieties (Gordon 1999); CD14 can bind ICAM3 (Gregory et al. 1998); and CD91 (or LRP1) binds C1q via calreticulin (Gardai et al. 2005). Here, we will focus on the PS recognition system by phagocytes as a model for dead cell recognition and clearance.

Despite a common goal, PS receptors differ in their expression patterns, mode of PS recognition, and downstream signaling. In addition to professional phagocytes, these receptors are expressed in a variety of tissues, including bone marrow (BAI-1), spleen (BAI-1), brain (BAI-1), lungs (RAGE), kidney cells (TIM-1), and sinusoidal endothelial cells (stabilin-2) (Hochreiter-Hufford and Ravichandran

2013). The tissue specificity of these receptors may help to explain why multiple PS receptors are required for efficient efferocytosis, as different tissues may require specialized PS receptor mechanisms (Camins et al. 2008; Nagata et al. 2010; Poon et al. 2014). For example, defects in BAI-1, highly expressed in glial and neuronal cells, are associated with neurodegenerative disorders (Sokolowski and Mandell 2011), while stabilin-2 expression is highly expressed in endothelial cells within atherosclerotic plaques (Lee et al. 2011).

Likewise, recognition of PS by these molecules occurs via different domains. The TIM family of receptors utilizes an IgV domain (Santiago et al. 2007). BAI-1 binds PS via thrombospondin type 1 repeats (Park et al. 2007), while stabilin-2 contains EGF-like domains that mediate PS recognition (Park et al. 2008a). MFG-E8 binds PS via C1 and C2 domains (Hanayama et al. 2002). Why different receptors or bridging molecules have evolved different PS-binding motifs is unknown and adds an additional layer of complexity to the study of efferocytosis.

Engagement of these PS receptors (or surface receptors engaged by bridging molecules) results in cytoskeletal reorganization to facilitate phagocytosis. Studies to delineate the molecules involved in the engulfment of dead cells in *C. elegans* and subsequent identification of mammalian homologues have begun to clarify the intracellular signaling events that occur (Reddien and Horvitz 2004). The uptake of dead cells is mediated by the Rho family of small GTPases, including members RhoA, Rac, Rab5, and Cdc42 (Nakaya et al. 2006), which cycle between the resting, inactive GDP-bound state and the active GTP-bound state, mediated by specific guanine nucleotide exchange factors (GEFs).

The physical engulfment of apoptotic cells is morphologically distinct from other types of phagocytosis. Whereas particles engulfed via complement-receptor-mediated phagocytosis seem to be absorbed into the phagocyte, dead cell engulfment involves active membrane ruffling by a process similar to macropinocytosis (Olazabal et al. 2002; Riento and Ridley 2003). Specific Rho family GTPases are conversely activated or inactivated during phagocytosis. Unlike its role in complement-receptor-mediated phagocytosis, it has been demonstrated that RhoA negatively regulates engulfment, as inhibition of this GTPase enhances engulfment. Conversely, overexpression of RhoA inhibits engulfment, and this is dependent on signaling by the RhoA-binding protein, Rho-associated coiled-coil-containing protein kinase (ROCK) (Erwig et al. 2006). An increase in the kinase activity of ROCK affects the status of myosin light chain (MLC) phosphorylation, to promote actomyosin assembly and cell contractility (Riento and Ridley 2003). This decrease in RhoA activation, and in turn the decreased signaling via ROCK, decreases stress fiber formation and probably facilitates cell shape changes during engulfment (Erwig et al. 2006).

Signaling during apoptotic cell engulfment converges on evolutionarily conserved pathway that leads to Rac1 activation, though the molecules used by specific receptors to activate the Rac1 pathway can differ. Unlike RhoA, Rac1 activation translates into membrane ruffles that are necessary to promote phagocytosis (Nakaya et al. 2006). Similarly, CDC42 has been linked to the engulfment of apoptotic cells, although its precise role is unclear (Leverrier et al. 2001)

Engagement of integrins, such as $\alpha_v\beta_3$ or $\alpha_v\beta_5$, or Mer by bridging molecules recruits the cytoplasmic protein, CrkII, which then associates with the adaptor proteins ELMO1 and DOCK180 to the phagocytic cup (Albert et al. 2000; Wu et al. 2005). Together, DOCK180 and ELMO1 form a bipartite GEF, which activates Rac1 (Brugnera et al. 2002). BAI1 also requires the activity of the DOCK180/ELMO1 complex for Rac1 activation, but BAI1 is able to recruit and bind ELMO1 independently (Park et al. 2007). The importance of DOCK180 and ELMO1 is highlighted by experiments in *C. elegans* wherein loss of the ELMO1 homologue (CED-12) results in failure to clear apoptotic cells during development (Wu et al. 2001). Moreover, overexpression of DOCK180 and ELMO1 results in greatly increased Rac1 activity and phagocytic capacity (Brugnera et al. 2002).

Not all "eat-me" signal receptors require the DOCK180/ELMO1 complex for Rac1 activation, though. Stabilin-2 requires the activity of the adaptor protein, GULP, to activate the Rac1 pathway (Park et al. 2008b, 2010b). Likewise, CD91/LRP, which binds calreticulin, interacts with GULP (Su et al. 2002). TIM4, however, contains a very short cytoplasmic region, which is not necessary for Rac1 activation, and currently, the signaling components downstream of TIM4 are unknown (Park et al. 2009).

Regardless of the path, Rac1 activation is a critical point in the engulfment process. GTP-bound Rac acts at sites of apoptotic cell recognition to promote Arp2/3 activation/actin polymerization/cytoskeletal rearrangement via the Scar/WAVE complex (Miki et al. 1998; Castellano et al. 2000). Once engulfment is complete, however, the job of phagocyte is not over. It has merely just begun.

3.4 Aperitifs and Digestifs: Digestion and Immune Response

Once encased inside of the phagocyte, the dead cell is now cargo destined for digestion and degradation. Late-stage RhoA activation is thought to promote apoptotic cell digestion by regulating the acidification of phagosomes (Erwig and Henson 2008). Similarly, GDP-bound small GTPase, Rab5, is recruited to the phagosome and activated by yet-unidentified GEFs. This GTP-bound, active Rab5 promotes activity of the Class III PI3 K, VPS34, which generates PI(3)P on the phagosomal membrane (Kinchen 2010). Recent studies connect Rab5 activation to Rab7 recruitment to the phagosome (Nordmann et al. 2010). Subsequently, the HOPS complex is recruited to the phagosome and activates Rab7, resulting in fusion of the phagosome with the lysosomal network (Kinchen et al. 2008).

Once the phagosome becomes mature via lysosomal fusion, acidic proteases and nucleases get activated and the apoptotic cell targets are degraded into their basic cellular components including fats, sterols, peptides, and nucleotides. DNAse II is a lysosomal enzyme required for the degradation of DNA, and DNAse II deficiency results in an accumulation of undigested DNA fragments within phagocytic cells, as well as polyarthritis and inflammation in joint tissues (Kawane et al. 2001, 2006).

But like any meal, it can have a profound effect on the eater. An area of growing interest is how a phagocyte handles the metabolic stress of ingesting a cellular corpse and essentially doubling its content of cellular components. One such component, cholesterol, can have a profound effect on the phagocyte's response to engulfed dead cells. In order to maintain their homeostasis in the face of increased cholesterol, phagocytes increase their basal cholesterol efflux activity from the cell (Noelia et al. 2009). Engagement of PS receptors leads to the activation of peroxisome proliferator-activated receptor γ/δ (PPARγ/δ) and liver x receptor (LXR) families, both important regulators of cellular lipid homeostasis and the clearance of apoptotic cells (Mukundan et al. 2009; Roszer et al. 2011). This induction results in the upregulation of phagocytic receptors, such as members of the TAM family, and basal cholesterol efflux machinery, such as 12-transmembrane protein ABCA1 (ATP-binding cassette subfamily A, member 1), to accommodate the increase in cholesterol associated with engulfment (Han and Ravichandran 2011). Moreover, $PPAR\gamma^{-/-}$ and $\mathrm{PPAR}\delta^{-/-}$ macrophages show a defect in apoptotic cell uptake. The dual functions of PPARs and LXRs in both lipid apoptotic cell clearance and lipid homeostasis suggest the interconnectedness between efferocytosis and metabolism.

Apoptotic cell death occurs in healthy organisms as part of normal tissue turnover and thus needs to be immunologically silent with regard to its resolution (Henson and Hume 2006). One of the hallmarks of apoptotic cell clearance is its non-inflammatory and non-immunogenic nature, and cholesterol homeostasis plays a critical role in this tolerant pathway (Hochreiter-Hufford and Ravichandran 2013; Poon et al. 2014). Phagocytes that have engulfed apoptotic cells have been shown to secrete anti-inflammatory cytokines, such as TGFβ and interleukin-10 (IL-10) (Fadok et al. 1998a, 2001). Moreover, the uptake of apoptotic cells actively suppressed pro-inflammatory cytokines, such as tumor necrosis factor (TNF), IL-1 and IL-12 (Kim et al. 2004). Intriguingly, PPARγ and PPARδ are central players in the polarization of M2 macrophages, the phenotype of which is anti-inflammatory. Agonists for both PPARγ and LXR have been shown to inhibit inflammatory responses (Mukundan et al. 2009; Noelia et al. 2009; Poon et al. 2014).

How then do apoptotic cells program their engulfers to tolerate their presence? Engulfment of necrotic cells or opsonization of corpse debris via FcR-mediated phagocytosis does not induce enhanced cholesterol efflux in the phagocytes, despite providing excess cholesterol for the engulfing cells (Kiss et al. 2006). These data suggest it is not the burden of extra cholesterol, but the engagement of ligands on apoptotic cells that induce a "prophylactic" cholesterol efflux from phagocytes. One key ligand seems to be the exposed PS on apoptotic cells. Coculture with mere apoptotic membranes or PS liposomes can induce the cholesterol efflux, anti-inflammatory cytokine production, and suppression of pro-inflammatory genes (Huynh et al. 2002; Kim et al. 2004). Collectively, these data suggest that recognition of apoptotic cells via engagement of "eat-me" signals, specifically PS, contributes to the silent clearance of these cells. Although PS recognition seems to be relevant for such immune tolerance, the roles of other "eat-me" signals and cognate receptor(s), as well as other signaling pathways involved, are unknown.

Despite engaging PS receptors, necrotic, necroptotic, or pyroptotic cells do not elicit an anti-inflammatory response (Martinez et al. 2011, 2013). Moreover, lytic cell death, such as necrosis, necroptosis, and pyroptosis, often results in an increased release of "find-me" signals (Ravichandran 2010). In the absence of efficient efferocytosis, apoptotic cells will undergo secondary necrosis and lyse, causing an inflammatory response (Juncadella et al. 2013). Of note, it is hypothesized that merely caspase activation results in dampening of the immunogenicity of DAMPS, and therefore, apoptotic cells that have transitioned through secondary necrosis are not inflammatory (Kazama et al. 2008; Luthi et al. 2009). However, recent data have shown that apoptotic cells can indeed be inflammatory if they are not phagocytosed in a timely manner (Obeid et al. 2007; Michaud et al. 2011).

How the phagocyte handles the ingested corpse in terms of its processing, degradation, and subsequent influence on the pursuant immune response is an area of growing interest. From the perspective of a single-celled organism, the two ancient systems of phagocytosis and autophagy represent two modes of nutrient acquisition—phagocytosis when extracellular fuel is abundant and autophagy when nutrients are scarce. However, these scenarios become decidedly more complex when one considers the engulfment of pathogens or dead cells (Martinez et al. 2013). The discovery of LC3-associated phagocytosis (LAP) has shed some light on this issue. LAP is a process that marries the evolutionarily conserved pathways of phagocytosis and autophagy into a fundamentally new concept, allowing us to reimagine the impact of the autophagy machinery on innate host defense mechanisms. LAP is triggered wherein an extracellular particle, such as a pathogen, immune complex, or dead cell, is sensed and phagocytosed, and this engulfment recruits some, but not all, members of the autophagy machinery to the cargo-containing, single-membraned vesicle (Sanjuan et al. 2007; Martinez et al. 2011). Engagement of multiple types of receptors, including TLR1/2, TLR2/6, TLR4, FcR, and TIM4 has been shown to induce translocation of autophagy machinery and ultimately LC3 to the cargo-containing phagosome, termed the LAPosome (Florey et al. 2011; Martinez et al. 2011; Henault et al. 2012; Martinez 2015). It is the activity of these autophagic players that facilitates the rapid destruction of the cargo via fusion with the lysosomal pathway. This is not macroautophagy, per se, but a distinct process, and it is triggered upon phagocytosis of particles containing ligands that engage a receptor-mediated signaling pathway.

Molecularly, LAP differs from canonical autophagy in multiple ways. LAP proceeds independently of the preinitiation complex, composed of ULK1/2, FIP200, and ATG13, whereas autophagy requires its activity (Florey et al. 2011; Martinez et al. 2011). Both LAP and autophagy utilize the Class III PI3 K complex and its core components Beclin-1 and VPS34, but LAP exclusively utilizes the UVRAG-containing Class III PI3 K complex (Martinez et al. 2015). In addition to the Class III PI3 K complex, LAP also requires Rubicon (RUN domain protein as

Beclin-1 interacting and cysteine-rich containing), which negatively regulates autophagy, via its inhibition of VPS34 (Matsunaga et al. 2009; Zhong et al. 2009) or by blocking GTPase Rab7 activation (Sun et al. 2010). Rubicon associates constitutively with the UVRAG-containing Class III PI3 K complex (Matsunaga et al. 2009), as well as the NOX2 complex, also required for LAP (Martinez et al. 2015). Rubicon is crucial for both the interaction of the Class III PI3 K complex with the LAPosome and subsequent PI(3)P production as well as promoting the production of ROS via the NOX2 complex (Yang et al. 2012; Martinez et al. 2015).

Importantly, LAP can have a profound effect on the immune response to the engulfed material. LAP deficiency results in a failure to efficiently degrade intra-phagosomal yeast in vitro (Sanjuan et al. 2007) or clear *Aspergillus fumigatus* in vitro or in vivo (Martinez et al. 2015). Furthermore, lungs and serum from animals with LAP deficiency display increased levels of pro-inflammatory cytokines when challenged intranasally with *Aspergillus fumigatus* (Martinez et al. 2015). Additionally, LAP is a critical regulator of the type I interferon response to immune complexes (IC) by plasmacytoid dendritic cells (pDC). Engagement of the FcγR by the IC induces LAP, resulting in LC3 translocation to the IC-containing phagosome in an ATG5- and ATG7-dependent, but ULK1-, FIP200-, and ATG13-independent manner. This failure to translocate LC3 results in a failure to acquire a late-endolysosomal phenotype, and subsequently a failure to form the specialized IRF7-signaling compartment required for TLR9-mediated activation of interferon regulatory factor 7 (IRF7). IFN-α production was completely ablated in $ATG7^{-/-}$, but not $ULK1^{-/-}$, pDC in response to DNA-IC, suggesting that LAP could affect the functional immune response elicited by this autoantigen (Henault et al. 2012).

LAP can act as a critical defense mechanism against autoimmune responses. Whereas much has been explored in terms the link between the uptake of dead cells with autoimmunity, how the phagocyte degrades the engulfed corpse is also a critical component to preventing unwanted immune responses. Billions of cells die daily as a result of stress, infection, or normal homeostasis, and it is the responsibility of the phagocytes of the immune system, such as macrophages, to rid the body of these cellular corpses, thus preventing inflammation and autoimmunity (Han and Ravichandran 2011; Martinez et al. 2011). Importantly, LAP has been demonstrated to play a critical role in the efficient clearance of dying cells, as well as promoting the anti-inflammatory response to apoptotic cells. Engagement of the PS receptor, TIM4, results in recruitment of the autophagic machinery to the dead-cell-containing, single-membrane phagosome. Macrophages deficient for ATG7, but not ULK1, fail to recruit LC3 to the phagosome, which results in failures in phagosomal acidification and subsequent corpse degradation. Whereas the phagocytosis of apoptotic cells is generally considered an "immunologically silent" event, ATG7-deficient macrophages produce dramatically increased levels of IL-1β and IL-6 when fed apoptotic cells. Moreover, these ATG7-deficient macrophages produce significantly less anti-inflammatory cytokines, such as IL-10, upon such engulfment (Martinez et al. 2011). How the LAP pathway modulates the immune response to apoptotic cells remains to be elucidated. The process of phagosome maturation and its effect on the immune response are actively being studied, in order to further understand the role of dead cell clearance in the prevention of autoimmunity and other pathologies (discussed below).

4 Food Poisoning: Pathologies Associated with Aberrant Efferocytosis

The lack of detectable apoptotic cells under physiological conditions speaks to the sheer efficiency of efferocytosis. Conversely, but not surprisingly, many different diseases are characterized by the presence of uncleared dead cells or a defect in properly handling engulfed cellular corpses. Furthermore, the unwanted inflammatory response to uncleared dead cells can exacerbate some conditions (Hochreiter-Hufford and Ravichandran 2013; Poon et al. 2014). Non-resolving inflammation contributes to tissue damage and organ dysfunction in a wide array of pathologies. Accumulating evidence indicates that efferocytosis is impaired in many autoimmune and inflammatory disorders.

4.1 Systemic Lupus Erythematosus (SLE)

Perhaps the most common pathological disorder associated with aberrant dead cell clearance is systemic lupus erythematosus (SLE). SLE is a chronic systemic autoimmune disorder affecting the skin, lungs, kidneys, and central nervous system. Patients with SLE display persistence of apoptotic cells within lymph nodes, blood, and skin (Baumann et al. 2002). As discussed earlier, apoptotic cells, while regarded immunologically silent initially, can undergo secondary necrosis if cleared in efficiently. This results in rupture of the protective plasma membrane and the release of intracellular autoantigens, such as DNA, ATP, and HMGB1, normally compartmentalized within an apoptotic cell (Raffray and Cohen 1997; Peter et al. 2010; Nikoletopoulou et al. 2013). SLE patients show a strong correlation between disease progression and the failed clearance of apoptotic cells, as well as increased inflammation (Liu and Davidson 2012). SLE patients also contain elevated levels of DNA or nucleosomes in the circulation, as well as autoantibodies that are specific for nuclear or other "self" components (Rumore and Steinman 1990). These autoantibodies can bind to circulating autoantigens, forming immune complexes that accumulate or deposit in the glomerular and vessel walls of the kidney, causing lupus nephritis (Berden 1997; van Bruggen et al. 1997).

It has been demonstrated that mice with deficiencies for engulfment (and hence clearance) of apoptotic cells, such as MFGE8-, BAI1-, TIM4-, or MerTK-deficient animals, accumulate apoptotic corpses within lymph nodes and develop an SLE-like disease that involves autoantibody formation, splenomegaly, and glomerulonephritis

(Cohen et al. 2002; Hanayama et al. 2004; Park et al. 2007; Rodriguez-Manzanet et al. 2010). Furthermore, genetic polymorphisms and aberrant splicing of MFGE8 have been reported in a small subset of patients with SLE, indicating that this pathway of apoptotic cell recognition and clearance could be deregulated in some patients (Hu et al. 2009). Similarly, deficiencies in components of the complement pathway, particularly C1q which plays a key role in apoptotic cell clearance, have been also associated with SLE (Botto et al. 1998).

Not only is phagocytic capacity critical for the prevention of SLE, but the efficient degradation of engulfed dead cells is also an important factor. Mice deficient for DNAse I, critical for the degradation of chromatin, are prone to SLE-like glomerulonephritis (Napirei et al. 2000). Intriguingly, genome-wide association studies have identified polymorphisms in *atg5* (Zhou et al. 2011) and possibly *atg7* (Clarke et al. 2014), genes involved in both canonical autophagy and LAP (Mizushima 2007; Florey et al. 2011; Martinez et al. 2011, 2015; Henault et al. 2012; Kim et al. 2013), as predisposition markers for SLE.

4.2 Rheumatoid Arthritis

Rheumatoid arthritis is a chronic autoimmune disease associated with progressive joint destruction. It is a systemic inflammatory disorder characterized by increased circulating autoantibodies against citrullinated peptides or the complement protein, C3 (Luban and Li 2010; Kenyon et al. 2011). While there exists little direct evidence that rheumatoid arthritis is caused by defects in efferocytosis, site of inflammation often contain DAMPS, such as HMGB1 or histones H3 and H4, characteristic of uncleared apoptotic cells (Friggeri et al. 2012). These components can bind to phagocytes and inhibit efferocytosis, therefore perpetuating progression to secondary necrosis and the release of immunostimulatory materials (Hurst et al. 1983; Friggeri et al. 2010). Mice deficient for DNAse II, and thus defective for the degradation of engulfed cellular cargo, develop polyarthritis and anemia associated with significant increased inflammatory markers, reminiscent of human rheumatoid arthritis (Kawane et al. 2006) While direct genetic links to efferocytosis have not been described for human rheumatoid arthritis, studies have demonstrated that increasing the levels of bridging molecules for TAM receptor or activating the LXR/PPARγ can have therapeutic benefits in mouse models of inflammatory arthritis (Park et al. 2010a).

4.3 Type 1 Diabetes

Type 1 diabetes (T1D) is a T cell-mediated autoimmune disease that results from destruction of the insulin-producing β-cells in the islets of Langerhans in the pancreas. The clearance of apoptotic cells (a source of self-antigens) by phagocytes, predominantly dendritic cells, induces a tolerogenic response. However, defective clearance of cellular corpses, however, can result in increased rate of β-cells apoptosis, inflammation, and loss of tolerance, as inefficiently cleared apoptotic cells undergo necrosis, releasing danger signals and autoantigens into the extracellular milieu (Vives-Pi et al. 2015).

One of the key factors in T1D is the loss of T cell tolerance to self-antigens. While the mechanisms by which efferocytosis induces selective immunosuppression are not fully understood, recent evidence indicates that impaired clearance of dying cells contributes to immunogenic, not tolerogenic, DC maturation and chronic inflammation. Previous work has demonstrated that macrophages from non-obese diabetic (NOD) mice, which spontaneously develop type 1 diabetes mellitus, have a profound defect in the phagocytosis of apoptotic cells in vitro. NOD mice also display impaired efferocytosis in vivo when challenged with apoptotic stimuli. This defect in apoptotic cell clearance by NOD mice also translated into increased production of antinuclear autoantibodies (ANA) (O'Brien et al. 2006). Another study demonstrated that the tolerogenic behavior of dendritic cells after islet cells efferocytosis is central to silencing autoreactive T cells in type I diabetes (Pujol-Autonell et al. 2013).

Impaired wound healing is a common complication encountered by patients with both type 1 and type 2 diabetes mellitus. A large number of neutrophils are recruited to the wound site but must be cleared adequately by macrophages to initiate the next stage of wound healing. Macrophages isolated from diabetic wounds display dysfunctional efferocytosis, resulting in increased dead cell burden at the wound site, increased inflammation, and delayed wound healing (Maruyama et al. 2007; Khanna et al. 2010). While it has been demonstrated that diabetes is associated with compromised efferocytosis and high levels of pro-inflammatory cytokines, the mechanisms underlying these defects have not been characterized.

4.4 Atherosclerosis

Atherosclerosis is associated with chronic inflammation of the vascular wall, predominantly caused by the recruitment and accumulation of monocytes, macrophages, and dendritic cells (DCs). These phagocytic cells engulf oxidized lipids, causing the lipid-laden cells to undergo apoptosis, wherein they themselves become engulfed by neighboring phagocytes (Tabas 2005; Ley et al. 2011). While efficient efferocytosis can compensate during the initial onset of atherosclerosis, mature atherosclerotic lesions (known as plaques) are characterized by the presence of foam cells, macrophages that have taken up necrotic cells at the core of plaques and failed to stimulate cholesterol efflux (Vucic et al. 2012; Poon et al. 2014). Pathologically, this results in further reduced clearance of apoptotic cells, secondary necrosis, expansion of plaque lesions, and eventual plaque rupture, leading directly to acute coronary syndromes and stroke in humans (Schrijvers et al. 2005; Tabas 2005).

Mouse models with defects in the efferocytosis machinery, such as MerTK-, MFG-E8-, or C1q-deficient models, display an accumulation of apoptotic cellular debris within plaques and exacerbated atherosclerosis (Tabas 2005; Bhatia et al. 2007; Thorp et al. 2008). Apoptotic cell uptake via interaction with ICAM3 can be bound and inhibited by oxidized lipids present in plaques (Miller et al. 2003).

While the correlation between decreased efferocytosis and atherosclerotic lesions is not incompletely understood, several clinical observations linked to possible mechanisms have been described. HMG-CoA reductase inhibitors (also known as statins) are commonly used pharmacological agents for the treatment of atherosclerosis and vascular disease. In addition to lowering cholesterol and inflammation, an additional mechanism of action of statins is the inhibition of RhoA, a negative regulator of engulfment that is highly expressed in atherosclerotic lesions (Loirand et al. 2006).

Downstream of engulfment, activation of the nuclear receptors liver X receptors (LXRs) and peroxisome proliferator-activated receptors (PPARs) are critical to promoting tolerance, as well as by upregulating MerTK expression. Moreover, synthetic agonists to LXR or PPARs have been demonstrated to be beneficial to the treatment of atherosclerosis (Noelia et al. 2009). Intricately linked to the pathological progression of atherosclerosis is cholesterol efflux. Uptake of apoptotic cells by phagocytes stimulates cholesterol efflux, primarily though upregulation of ABCA1, a critical molecule that transports free cholesterol from within the cells to lipid-poor apoA1 that is then modified in the plasma for transport to the liver and excretion (Oram and Heinecke 2005; Cuchel and Rader 2006). Indeed, decreased ABCA1 activity in mouse models promotes inflammation and atherosclerotic lesions. Conversely, overexpression of ABCA1 can dampen the inflammatory response and reverses the disease (Tang et al. 2009; Zhu et al. 2010). Finally, mice with deficiencies in the autophagic machinery, shared by the LAP pathway, such as LysM-Cre Atg5flox/flox mice, display more atherosclerotic lesions, suggesting the possibility that LAP-mediated processing is required for the prevention of inflammation (Liao et al. 2012; Razani et al. 2012).

4.5 Lung Inflammation

Upon the induction of lung inflammation, neutrophils, the most abundant cell type involved in the innate immune response, are quickly recruited airway to airway. Neutrophils, however, as a short-lived population, undergo apoptosis in order to prevent the release of histotoxic contents and subsequent damage to surrounding tissues. Clearance of these apoptotic neutrophils by phagocytes is central to the successful resolution of the inflammatory response (Fox et al. 2010). Both mouse and human models of chronic obstructive pulmonary disease and cystic fibrosis have demonstrated that impaired efferocytosis of dying neutrophils during inflammation can lead to a prolonged inflammatory response. Chronic obstructive pulmonary disease (COPD) is a common, yet complex disease, highly associated with cigarette smoke. Chronic inflammation, extracellular matrix destruction, and increased airway epithelial cell and neutrophil apoptosis are reported in COPD models, and alveolar macrophages from COPD patients demonstrate a decreased phagocytic capacity for apoptotic cells (Hodge et al. 2003). Cystic fibrosis (CF) lung disease is characterized by early, protracted inflammation associated with a massive influx of neutrophils and other inflammatory cells, and the efficient clearance of these inflammatory cells (now apoptotic) by phagocytes is critical to the resolution of inflammation. Similar to COPD patients, alveolar macrophages from CF patients display defective efferocytosis, possibly attributable to necrotic neutrophil-derived proteases capable of cleaving PS receptors on the surface of phagocytes (Vandivier et al. 2002a, b).

Human allergic asthma is a chronic inflammatory disorder of the airways and is characterized by airway inflammation, persistent airway hyperresponsiveness (AHR), and intermittent, reversible airway obstruction. "Airway remodeling," including airway fibrosis, goblet cell hyperplasia, and other structural changes, is thought to be the result of chronic inflammation and serves to exacerbate the asthma symptoms (Nials and Uddin 2008). Alveolar macrophages from patients with sever asthma or poorly controlled asthma are defective in clearing apoptotic cells (Huynh et al. 2005; Fitzpatrick et al. 2008). Similarly, patients with non-eosinophilic asthma have increased numbers of neutrophils in the airways, and alveolar macrophages from these patients show an impaired ability to phagocytose apoptotic cells (Simpson et al. 2013).

While the exact mechanisms underlying defective phagocytosis in patients with severe asthma are not yet understood, the use of corticosteroids, the most common treatment in asthma, induces eosinophil apoptosis as well as eosinophil engulfment by macrophages in vitro, via the binding of protein S to apoptotic eosinophils and the upregulation of MerTK on the surface of macrophages (Liu et al. 1999; McColl et al. 2009). Unsurprisingly, this treatment is significantly less effective in non-eosinophilic asthma and other neutrophil-dominated lung inflammatory disorders (Vago et al. 2012; Poon et al. 2014).

4.6 Neurodegenerative Disorders

In the peripheral immune system, phagocytes tasked with efferocytosis are mainly comprised of macrophages, monocytes, and dendritic cells. In the brain, however, microglia act as resident macrophages to accomplish this function. Microglia have phenotypical similarities to peripheral macrophages, in that they express and utilize PS receptors for the recognition and uptake of dead cells (Witting et al. 2000). Multiple neurodegenerative disorders have been associated with defective efferocytosis, such as Alzheimer's disease, Parkinson's disease, and Huntington's disease (Mattson 2000). The most direct link of efferocytosis to neurodegenerative diseases is MFG-E8. Microglia express MFG-E8, and treatment of microglia cultures with fractalkine (CX3CL1) increases MFG-E8 mRNA levels (Leonardi-Essmann et al.

2005; Fuller and Van Eldik 2008). Inhibition of MFG-E8 in microglia cells resulted in decreased engulfment of apoptotic neurons (Fuller and Van Eldik 2008). Of note, it has also been reported that MFG-E8 also mediates the phagocytosis of viable, LPS-activated neurons, resulting in death of the engulfed neuron (Fuller and Van Eldik 2008; Fricker et al. 2012). Finally, MFG-E8 levels in the brains of the Tg2576 mouse model of Alzheimer's disease were severely decreased with age compared to wild-type controls (Fuller and Van Eldik 2008). Taken together, MFG-E8 seems to play a central role in the clearance of dead cells in the brain, and deficiencies in MFG-E8 can contribute to the onset and severity of Alzheimer's disease.

Other proteins that have tangential roles in efferocytosis have been implicated in neurodegenerative disorders. Apolipoprotein E (apoE) is a cholesterol transport protein expressed in liver, central nervous system, vascular smooth muscle cells, adrenals, macrophages, and adipocytes. The E4 isoform of apoE (apoE4) is a major genetic risk factor for multiple inflammatory metabolic diseases, including atherosclerosis and Alzheimer's disease. Peritoneal macrophages isolated from *APOE4* mice demonstrate defects in efferocytosis and increased inflammatory response, presumably through ER stress (Cash et al. 2012). How the cells of the brain mediate the clearance of potentially damaging dead cells, as well as other debris, is currently an area of great interest.

5 Conclusions

The controlled cell death program of apoptosis is an integral part of maintaining development and cell turnover, yet like a large meal, too much of a good thing can be detrimental. The sheer magnitude of the task undertaken by professional phagocytes to keep an organism free of potentially dangerous dead cells, and thus unwanted autoimmunity and inflammation, is a daunting one, but one performed with exquisite accuracy under normal conditions. Indeed, many autoimmune conditions have been clearly linked to defects in efferocytosis, in terms of recognition, engulfment, and digestion of its cellular corpse cargo. While the field of efferocytosis is relatively young, these new insights into the underlying mechanisms of dead cell clearance will provide invaluable opportunities to attack autoimmune and autoinflammatory diseases at its core. Bon appetit!

References

Albert ML, Kim JI, Birge RB (2000) alphavbeta5 integrin recruits the CrkII-Dock180-rac1 complex for phagocytosis of apoptotic cells. Nat Cell Biol 2(12):899–905

Ashkenazi A, Dixit VM (1998) Death receptors: signaling and modulation. Science 281 (5381):1305–1308

Balasubramanian K, Schroit AJ (2003) Aminophospholipid asymmetry: a matter of life and death. Annu Rev Physiol 65:701–734

Baumann I, Kolowos W, Voll RE, Manger B, Gaipl U, Neuhuber WL, Kirchner T, Kalden JR, Herrmann M (2002) Impaired uptake of apoptotic cells into tingible body macrophages in germinal centers of patients with systemic lupus erythematosus. Arthritis Rheum 46(1):191–201

Berden JH (1997) Lupus nephritis. Kidney Int 52(2):538–558

Bergsbaken T, Fink SL, Cookson BT (2009) Pyroptosis: host cell death and inflammation. Nat Rev Microbiol 7(2):99–109

Bhatia VK, Yun S, Leung V, Grimsditch DC, Benson GM, Botto MB, Boyle JJ, Haskard DO (2007) Complement C1q reduces early atherosclerosis in low-density lipoprotein receptor-deficient mice. Am J Pathol 170(1):416–426

Bidere N, Su HC, Lenardo MJ (2006) Genetic disorders of programmed cell death in the immune system. Annu Rev Immunol 24:321–352

Boatright KM, Renatus M, Scott FL, Sperandio S, Shin H, Pedersen IM, Ricci JE, Edris WA, Sutherlin DP, Green DR, Salvesen GS (2003) A unified model for apical caspase activation. Mol Cell 11(2):529–541

Botto M, Dell'Agnola C, Bygrave AE, Thompson EM, Cook HT, Petry F, Loos M, Pandolfi PP, Walport MJ (1998) Homozygous C1q deficiency causes glomerulonephritis associated with multiple apoptotic bodies. Nat Genet 19(1):56–59

Bournazou I, Pound JD, Duffin R, Bournazos S, Melville LA, Brown SB, Rossi AG, Gregory CD (2009) Apoptotic human cells inhibit migration of granulocytes via release of lactoferrin. J Clin Invest 119(1):20–32

Bournazou I, Mackenzie KJ, Duffin R, Rossi AG, Gregory CD (2010) Inhibition of eosinophil migration by lactoferrin. Immunol Cell Biol 88(2):220–223

Bratton DL, Henson PM (2008) Apoptotic cell recognition: will the real phosphatidylserine receptor(s) please stand up? Curr Biol 18(2):R76–R79

Brenner D, Mak TW (2009) Mitochondrial cell death effectors. Curr Opin Cell Biol 21(6):871–877

Brugnera E, Haney L, Grimsley C, Lu M, Walk SF, Tosello-Trampont AC, Macara IG, Madhani H, Fink GR, Ravichandran KS (2002) Unconventional Rac-GEF activity is mediated through the Dock180-ELMO complex. Nat Cell Biol 4(8):574–582

Camins A, Pallas M, Silvestre JS (2008) Apoptotic mechanisms involved in neurodegenerative diseases: experimental and therapeutic approaches. Methods Find Exp Clin Pharmacol 30 (1):43–65

Cash JG, Kuhel DG, Basford JE, Jaeschke A, Chatterjee TK, Weintraub NL, Hui DY (2012) Apolipoprotein E4 impairs macrophage efferocytosis and potentiates apoptosis by accelerating endoplasmic reticulum stress. J Biol Chem 287(33):27876–27884

Castellano F, Montcourrier P, Chavrier P (2000) Membrane recruitment of Rac1 triggers phagocytosis. J Cell Sci 113(Pt 17):2955–2961

Chekeni FB, Elliott MR, Sandilos JK, Walk SF, Kinchen JM, Lazarowski ER, Armstrong AJ, Penuela S, Laird DW, Salvesen GS, Isakson BE, Bayliss DA, Ravichandran KS (2010) Pannexin 1 channels mediate 'find-me' signal release and membrane permeability during apoptosis. Nature 467(7317):863–867

Clarke AJ, Ellinghaus U, Cortini A, Stranks A, Simon AK, Botto M, Vyse TJ (2014) Autophagy is activated in systemic lupus erythematosus and required for plasmablast development. Ann Rheum Dis

Cohen PL, Caricchio R, Abraham V, Camenisch TD, Jennette JC, Roubey RA, Earp HS, Matsushima G, Reap EA (2002) Delayed apoptotic cell clearance and lupus-like autoimmunity in mice lacking the c-mer membrane tyrosine kinase. J Exp Med 196(1):135–140

Colussi PA, Kumar S (1999) Targeted disruption of caspase genes in mice: what they tell us about the functions of individual caspases in apoptosis. Immunol Cell Biol 77(1):58–63

Cuchel M, Rader DJ (2006) Macrophage reverse cholesterol transport: key to the regression of atherosclerosis? Circulation 113(21):2548–2555

Dillon CP, Weinlich R, Rodriguez DA, Cripps JG, Quarato G, Gurung P, Verbist KC, Brewer TL, Llambi F, Gong YN, Janke LJ, Kelliher MA, Kanneganti TD, Green DR (2014) RIPK1 blocks early postnatal lethality mediated by caspase-8 and RIPK3. Cell 157(5):1189–1202

Elliott MR, Chekeni FB, Trampont PC, Lazarowski ER, Kadl A, Walk SF, Park D, Woodson RI, Ostankovich M, Sharma P, Lysiak JJ, Harden TK, Leitinger N, Ravichandran KS (2009) Nucleotides released by apoptotic cells act as a find-me signal to promote phagocytic clearance. Nature 461(7261):282–286

Elward K, Griffiths M, Mizuno M, Harris CL, Neal JW, Morgan BP, Gasque P (2005) CD46 plays a key role in tailoring innate immune recognition of apoptotic and necrotic cells. J Biol Chem 280(43):36342–36354

Erwig LP, Henson PM (2008) Clearance of apoptotic cells by phagocytes. Cell Death Differ 15 (2):243–250

Erwig LP, McPhilips KA, Wynes MW, Ivetic A, Ridley AJ, Henson PM (2006) Differential regulation of phagosome maturation in macrophages and dendritic cells mediated by Rho GTPases and ezrin-radixin-moesin (ERM) proteins. Proc Natl Acad Sci U S A 103(34):12825–12830

Ezekowitz RA, Sastry K, Bailly P, Warner A (1990) Molecular characterization of the human macrophage mannose receptor: demonstration of multiple carbohydrate recognition-like domains and phagocytosis of yeasts in Cos-1 cells. J Exp Med 172(6):1785–1794

Fadok VA, Bratton DL, Konowal A, Freed PW, Westcott JY, Henson PM (1998a) Macrophages that have ingested apoptotic cells in vitro inhibit proinflammatory cytokine production through autocrine/paracrine mechanisms involving TGF-beta, PGE2, and PAF. J Clin Invest 101 (4):890–898

Fadok VA, Warner ML, Bratton DL, Henson PM (1998b) CD36 is required for phagocytosis of apoptotic cells by human macrophages that use either a phosphatidylserine receptor or the vitronectin receptor (alpha v beta 3). J Immunol 161(11):6250–6257

Fadok VA, Bratton DL, Guthrie L, Henson PM (2001) Differential effects of apoptotic versus lysed cells on macrophage production of cytokines: role of proteases. J Immunol 166(11):6847–6854

Feoktistova M, Geserick P, Kellert B, Dimitrova DP, Langlais C, Hupe M, Cain K, MacFarlane M, Hacker G, Leverkus M (2011) cIAPs block Ripoptosome formation, a RIP1/caspase-8 containing intracellular cell death complex differentially regulated by cFLIP isoforms. Mol Cell 43(3):449–463

Fitzpatrick AM, Holguin F, Teague WG, Brown LA (2008) Alveolar macrophage phagocytosis is impaired in children with poorly controlled asthma. J Allergy Clin Immunol 121(6):1371–1378

Florey O, Kim SE, Sandoval CP, Haynes CM, Overholtzer M (2011) Autophagy machinery mediates macroendocytic processing and entotic cell death by targeting single membranes. Nat Cell Biol 13(11):1335–1343

Fox S, Leitch AE, Duffin R, Haslett C, Rossi AG (2010) Neutrophil apoptosis: relevance to the innate immune response and inflammatory disease. J Innate Immun 2(3):216–227

Freeman GJ, Casasnovas JM, Umetsu DT, DeKruyff RH (2010) TIM genes: a family of cell surface phosphatidylserine receptors that regulate innate and adaptive immunity. Immunol Rev 235(1):172–189

Fricker M, Neher JJ, Zhao JW, Thery C, Tolkovsky AM, Brown GC (2012) MFG-E8 mediates primary phagocytosis of viable neurons during neuroinflammation. J Neurosci 32(8):2657–2666

Friggeri A, Yang Y, Banerjee S, Park YJ, Liu G, Abraham E (2010) HMGB1 inhibits macrophage activity in efferocytosis through binding to the alphavbeta3-integrin. Am J Physiol Cell Physiol 299(6):C1267–C1276

Friggeri A, Banerjee S, Xie N, Cui H, De Freitas A, Zerfaoui M, Dupont H, Abraham E, Liu G (2012) Extracellular histones inhibit efferocytosis. Mol Med 18:825–833

Fuller AD, Van Eldik LJ (2008) MFG-E8 regulates microglial phagocytosis of apoptotic neurons. J Neuroimmune Pharmacol 3(4):246–256

Gardai SJ, McPhillips KA, Frasch SC, Janssen WJ, Starefeldt A, Murphy-Ullrich JE, Bratton DL, Oldenborg PA, Michalak M, Henson PM (2005) Cell-surface calreticulin initiates clearance of viable or apoptotic cells through trans-activation of LRP on the phagocyte. Cell 123(2):321–334
Gordon S (1999) Macrophage-restricted molecules: role in differentiation and activation. Immunol Lett 65(1–2):5–8
Green DR (2011) Means to an end : apoptosis and other cell death mechanisms. Cold Spring Harbor Laboratory Press, Cold Spring Harbor
Green DR, Ferguson T, Zitvogel L, Kroemer G (2009) Immunogenic and tolerogenic cell death. Nat Rev Immunol 9(5):353–363
Gregory CD, Devitt A, Moffatt O (1998) Roles of ICAM-3 and CD14 in the recognition and phagocytosis of apoptotic cells by macrophages. Biochem Soc Trans 26(4):644–649
Gude DR, Alvarez SE, Paugh SW, Mitra P, Yu J, Griffiths R, Barbour SE, Milstien S, Spiegel S (2008) Apoptosis induces expression of sphingosine kinase 1 to release sphingosine-1-phosphate as a "come-and-get-me" signal. FASEB J 22(8):2629–2638
Han CZ, Ravichandran KS (2011) Metabolic connections during apoptotic cell engulfment. Cell 147(7):1442–1445
Hanayama R, Tanaka M, Miwa K, Shinohara A, Iwamatsu A, Nagata S (2002) Identification of a factor that links apoptotic cells to phagocytes. Nature 417(6885):182–187
Hanayama R, Tanaka M, Miyasaka K, Aozasa K, Koike M, Uchiyama Y, Nagata S (2004) Autoimmune disease and impaired uptake of apoptotic cells in MFG-E8-deficient mice. Science 304(5674):1147–1150
Hart SP, Dransfield I, Rossi AG (2008) Phagocytosis of apoptotic cells. Methods 44(3):280–285
He M, Kubo H, Morimoto K, Fujino N, Suzuki T, Takahasi T, Yamada M, Yamaya M, Maekawa T, Yamamoto Y, Yamamoto H (2011) Receptor for advanced glycation end products binds to phosphatidylserine and assists in the clearance of apoptotic cells. EMBO Rep 12(4):358–364
Henao-Mejia J, Elinav E, Strowig T, Flavell RA (2012) Inflammasomes: far beyond inflammation. Nat Immunol 13(4):321–324
Henault J, Martinez J, Riggs JM, Tian J, Mehta P, Clarke L, Sasai M, Latz E, Brinkmann MM, Iwasaki A, Coyle AJ, Kolbeck R, Green DR, Sanjuan MA (2012) Noncanonical autophagy is required for type I interferon secretion in response to DNA-immune complexes. Immunity 37 (6):986–997
Henson PM, Hume DA (2006) Apoptotic cell removal in development and tissue homeostasis. Trends Immunol 27(5):244–250
Hochreiter-Hufford A, Ravichandran KS (2013) Clearing the dead: apoptotic cell sensing, recognition, engulfment, and digestion. Cold Spring Harb Perspect Biol 5(1):a008748
Hodge S, Hodge G, Scicchitano R, Reynolds PN, Holmes M (2003) Alveolar macrophages from subjects with chronic obstructive pulmonary disease are deficient in their ability to phagocytose apoptotic airway epithelial cells. Immunol Cell Biol 81(4):289–296
Hoeppner DJ, Hengartner MO, Schnabel R (2001) Engulfment genes cooperate with ced-3 to promote cell death in *Caenorhabditis elegans*. Nature 412(6843):202–206
Hu CY, Wu CS, Tsai HF, Chang SK, Tsai WI, Hsu PN (2009) Genetic polymorphism in milk fat globule-EGF factor 8 (MFG-E8) is associated with systemic lupus erythematosus in human. Lupus 18(8):676–681
Hurst NP, Nuki G, Wallington T (1983) Functional defects of monocyte C3b receptor-mediated phagocytosis in rheumatoid arthritis (RA): evidence for an association with the appearance of a circulating population of non-specific esterase-negative mononuclear phagocytes. Ann Rheum Dis 42(5):487–493
Huynh ML, Fadok VA, Henson PM (2002) Phosphatidylserine-dependent ingestion of apoptotic cells promotes TGF-beta1 secretion and the resolution of inflammation. J Clin Invest 109 (1):41–50
Huynh ML, Malcolm KC, Kotaru C, Tilstra JA, Westcott JY, Fadok VA, Wenzel SE (2005) Defective apoptotic cell phagocytosis attenuates prostaglandin E2 and 15-hydroxyeicosatetraenoic acid in severe asthma alveolar macrophages. Am J Respir Crit Care Med 172(8):972–979

Ishimoto Y, Ohashi K, Mizuno K, Nakano T (2000) Promotion of the uptake of PS liposomes and apoptotic cells by a product of growth arrest-specific gene, gas6. J Biochem 127(3):411–417
Iyer SS, Pulskens WP, Sadler JJ, Butter LM, Teske GJ, Ulland TK, Eisenbarth SC, Florquin S, Flavell RA, Leemans JC, Sutterwala FS (2009) Necrotic cells trigger a sterile inflammatory response through the Nlrp3 inflammasome. Proc Natl Acad Sci U S A 106(48):20388–20393
Jorgensen I, Miao EA (2015) Pyroptotic cell death defends against intracellular pathogens. Immunol Rev 265(1):130–142
Juncadella IJ, Kadl A, Sharma AK, Shim YM, Hochreiter-Hufford A, Borish L, Ravichandran KS (2013) Apoptotic cell clearance by bronchial epithelial cells critically influences airway inflammation. Nature 493(7433):547–551
Kaiser WJ, Upton JW, Long AB, Livingston-Rosanoff D, Daley-Bauer LP, Hakem R, Caspary T, Mocarski ES (2011) RIP3 mediates the embryonic lethality of caspase-8-deficient mice. Nature 471(7338):368–372
Kaiser WJ, Sridharan H, Huang C, Mandal P, Upton JW, Gough PJ, Sehon CA, Marquis RW, Bertin J, Mocarski ES (2013) Toll-like receptor 3-mediated necrosis via TRIF, RIP3, and MLKL. J Biol Chem 288(43):31268–31279
Kawane K, Fukuyama H, Kondoh G, Takeda J, Ohsawa Y, Uchiyama Y, Nagata S (2001) Requirement of DNase II for definitive erythropoiesis in the mouse fetal liver. Science 292 (5521):1546–1549
Kawane K, Ohtani M, Miwa K, Kizawa T, Kanbara Y, Yoshioka Y, Yoshikawa H, Nagata S (2006) Chronic polyarthritis caused by mammalian DNA that escapes from degradation in macrophages. Nature 443(7114):998–1002
Kayagaki N, Warming S, Lamkanfi M, Vande Walle L, Louie S, Dong J, Newton K, Qu Y, Liu J, Heldens S, Zhang J, Lee WP, Roose-Girma M, Dixit VM (2011) Non-canonical inflammasome activation targets caspase-11. Nature 479(7371):117–121
Kazama H, Ricci JE, Herndon JM, Hoppe G, Green DR, Ferguson TA (2008) Induction of immunological tolerance by apoptotic cells requires caspase-dependent oxidation of high-mobility group box-1 protein. Immunity 29(1):21–32
Kenyon KD, Cole C, Crawford F, Kappler JW, Thurman JM, Bratton DL, Boackle SA, Henson PM (2011) IgG autoantibodies against deposited C3 inhibit macrophage-mediated apoptotic cell engulfment in systemic autoimmunity. J Immunol 187(5):2101–2111
Khanna S, Biswas S, Shang Y, Collard E, Azad A, Kauh C, Bhasker V, Gordillo GM, Sen CK, Roy S (2010) Macrophage dysfunction impairs resolution of inflammation in the wounds of diabetic mice. PLoS ONE 5(3):e9539
Kim S, Elkon KB, Ma X (2004) Transcriptional suppression of interleukin-12 gene expression following phagocytosis of apoptotic cells. Immunity 21(5):643–653
Kim JY, Zhao H, Martinez J, Doggett TA, Kolesnikov AV, Tang PH, Ablonczy Z, Chan CC, Zhou Z, Green DR, Ferguson TA (2013) Noncanonical autophagy promotes the visual cycle. Cell 154(2):365–376
Kinchen JM (2010) A model to die for: signaling to apoptotic cell removal in worm, fly and mouse. Apoptosis 15(9):998–1006
Kinchen JM, Doukoumetzidis K, Almendinger J, Stergiou L, Tosello-Trampont A, Sifri CD, Hengartner MO, Ravichandran KS (2008) A pathway for phagosome maturation during engulfment of apoptotic cells. Nat Cell Biol 10(5):556–566
Kiss RS, Elliott MR, Ma Z, Marcel YL, Ravichandran KS (2006) Apoptotic cells induce a phosphatidylserine-dependent homeostatic response from phagocytes. Curr Biol 16 (22):2252–2258
Kono H, Rock KL (2008) How dying cells alert the immune system to danger. Nat Rev Immunol 8 (4):279–289
Kono H, Chen CJ, Ontiveros F, Rock KL (2010) Uric acid promotes an acute inflammatory response to sterile cell death in mice. J Clin Invest 120(6):1939–1949
Krysko DV, Vandenabeele P (2010) Clearance of dead cells: mechanisms, immune responses and implication in the development of diseases. Apoptosis 15(9):995–997
Kumar S (2007) Caspase function in programmed cell death. Cell Death Differ 14(1):32–43

LaRock CN, Cookson BT (2013) Burning down the house: cellular actions during pyroptosis. PLoS Pathog 9(12):e1003793

Lauber K, Bohn E, Krober SM, Xiao YJ, Blumenthal SG, Lindemann RK, Marini P, Wiedig C, Zobywalski A, Baksh S, Xu Y, Autenrieth IB, Schulze-Osthoff K, Belka C, Stuhler G, Wesselborg S (2003) Apoptotic cells induce migration of phagocytes via caspase-3-mediated release of a lipid attraction signal. Cell 113(6):717–730

Lee GY, Kim JH, Oh GT, Lee BH, Kwon IC, Kim IS (2011) Molecular targeting of atherosclerotic plaques by a stabilin-2-specific peptide ligand. J Control Release 155(2):211–217

Leist M, Jaattela M (2001) Four deaths and a funeral: from caspases to alternative mechanisms. Nat Rev Mol Cell Biol 2(8):589–598

Lemke G, Rothlin CV (2008) Immunobiology of the TAM receptors. Nat Rev Immunol 8(5):327–336

Leonardi-Essmann F, Emig M, Kitamura Y, Spanagel R, Gebicke-Haerter PJ (2005) Fractalkine-upregulated milk-fat globule EGF factor-8 protein in cultured rat microglia. J Neuroimmunol 160(1–2):92–101

Leventis PA, Grinstein S (2010) The distribution and function of phosphatidylserine in cellular membranes. Annu Rev Biophys 39:407–427

Leverrier Y, Lorenzi R, Blundell MP, Brickell P, Kinnon C, Ridley AJ, Thrasher AJ (2001) Cutting edge: the Wiskott-Aldrich syndrome protein is required for efficient phagocytosis of apoptotic cells. J Immunol 166(8):4831–4834

Ley K, Miller YI, Hedrick CC (2011) Monocyte and macrophage dynamics during atherogenesis. Arterioscler Thromb Vasc Biol 31(7):1506–1516

Liao X, Sluimer JC, Wang Y, Subramanian M, Brown K, Pattison JS, Robbins J, Martinez J, Tabas I (2012) Macrophage autophagy plays a protective role in advanced atherosclerosis. Cell Metab 15(4):545–553

Linkermann A, Hackl MJ, Kunzendorf U, Walczak H, Krautwald S, Jevnikar AM (2013) Necroptosis in immunity and ischemia-reperfusion injury. Am J Transplant 13(11):2797–2804

Liu Z, Davidson A (2012) Taming lupus-a new understanding of pathogenesis is leading to clinical advances. Nat Med 18(6):871–882

Liu Y, Cousin JM, Hughes J, Van Damme J, Seckl JR, Haslett C, Dransfield I, Savill J, Rossi AG (1999) Glucocorticoids promote nonphlogistic phagocytosis of apoptotic leukocytes. J Immunol 162(6):3639–3646

Llambi F, Green DR (2011) Apoptosis and oncogenesis: give and take in the BCL-2 family. Curr Opin Genet Dev 21(1):12–20

Llambi F, Moldoveanu T, Tait SW, Bouchier-Hayes L, Temirov J, McCormick LL, Dillon CP, Green DR (2011) A unified model of mammalian BCL-2 protein family interactions at the mitochondria. Mol Cell 44(4):517–531

Loirand G, Guerin P, Pacaud P (2006) Rho kinases in cardiovascular physiology and pathophysiology. Circ Res 98(3):322–334

Luban S, Li ZG (2010) Citrullinated peptide and its relevance to rheumatoid arthritis: an update. Int J Rheum Dis 13(4):284–287

Luthi AU, Cullen SP, McNeela EA, Duriez PJ, Afonina IS, Sheridan C, Brumatti G, Taylor RC, Kersse K, Vandenabeele P, Lavelle EC, Martin SJ (2009) Suppression of interleukin-33 bioactivity through proteolysis by apoptotic caspases. Immunity 31(1):84–98

Martinez J, Almendinger J, Oberst A, Ness R, Dillon CP, Fitzgerald P, Hengartner MO, Green DR (2011) Microtubule-associated protein 1 light chain 3 alpha (LC3)-associated phagocytosis is required for the efficient clearance of dead cells. Proc Natl Acad Sci USA 108(42):17396–17401

Martinez J, Verbist K, Wang R, Green DR (2013) The relationship between metabolism and the autophagy machinery during the innate immune response. Cell Metab 17(6):895–900

Martinez J, Malireddi RK, Lu Q, Cunha LD, Pelletier S, Gingras S, Orchard R, Guan JL, Tan H, Peng J, Kanneganti TD, Virgin HW, Green DR (2015) Molecular characterization of LC3-associated phagocytosis (LAP) reveals distinct roles for Rubicon, NOX2, and autophagy proteins. Nature Cell Biol

Maruyama K, Asai J, Ii M, Thorne T, Losordo DW, D'Amore PA (2007) Decreased macrophage number and activation lead to reduced lymphatic vessel formation and contribute to impaired diabetic wound healing. Am J Pathol 170(4):1178–1191

Matsunaga K, Saitoh T, Tabata K, Omori H, Satoh T, Kurotori N, Maejima I, Shirahama-Noda K, Ichimura T, Isobe T, Akira S, Noda T, Yoshimori T (2009) Two Beclin 1-binding proteins, Atg14L and Rubicon, reciprocally regulate autophagy at different stages. Nat Cell Biol 11 (4):385–396

Mattson MP (2000) Apoptosis in neurodegenerative disorders. Nat Rev Mol Cell Biol 1 (2):120–129

McColl A, Bournazos S, Franz S, Perretti M, Morgan BP, Haslett C, Dransfield I (2009) Glucocorticoids induce protein S-dependent phagocytosis of apoptotic neutrophils by human macrophages. J Immunol 183(3):2167–2175

McIlwain DR, Berger T, Mak TW (2015) Caspase functions in cell death and disease. Cold Spring Harb Perspect Biol **7**(4)

Michaud M, Martins I, Sukkurwala AQ, Adjemian S, Ma Y, Pellegatti P, Shen S, Kepp O, Scoazec M, Mignot G, Rello-Varona S, Tailler M, Menger L, Vacchelli E, Galluzzi L, Ghiringhelli F, di Virgilio F, Zitvogel L, Kroemer G (2011) Autophagy-dependent anticancer immune responses induced by chemotherapeutic agents in mice. Science 334(6062):1573–1577

Micheau O, Lens S, Gaide O, Alevizopoulos K, Tschopp J (2001) NF-kappaB signals induce the expression of c-FLIP. Mol Cell Biol 21(16):5299–5305

Miki H, Suetsugu S, Takenawa T (1998) WAVE, a novel WASP-family protein involved in actin reorganization induced by Rac. EMBO J 17(23):6932–6941

Miller YI, Viriyakosol S, Binder CJ, Feramisco JR, Kirkland TN, Witztum JL (2003) Minimally modified LDL binds to CD14, induces macrophage spreading via TLR4/MD-2, and inhibits phagocytosis of apoptotic cells. J Biol Chem 278(3):1561–1568

Miyanishi M, Tada K, Koike M, Uchiyama Y, Kitamura T, Nagata S (2007) Identification of Tim4 as a phosphatidylserine receptor. Nature 450(7168):435–439

Mizushima N (2007) Autophagy: process and function. Genes Dev 21(22):2861–2873

Mukundan L, Odegaard JI, Morel CR, Heredia JE, Mwangi JW, Ricardo-Gonzalez RR, Goh YP, Eagle AR, Dunn SE, Awakuni JU, Nguyen KD, Steinman L, Michie SA, Chawla A (2009) PPAR-delta senses and orchestrates clearance of apoptotic cells to promote tolerance. Nat Med 15(11):1266–1272

Muzio M, Stockwell BR, Stennicke HR, Salvesen GS, Dixit VM (1998) An induced proximity model for caspase-8 activation. J Biol Chem 273(5):2926–2930

Nagata S, Hanayama R, Kawane K (2010) Autoimmunity and the clearance of dead cells. Cell 140 (5):619–630

Nakaya M, Tanaka M, Okabe Y, Hanayama R, Nagata S (2006) Opposite effects of rho family GTPases on engulfment of apoptotic cells by macrophages. J Biol Chem 281(13):8836–8842

Napirei M, Karsunky H, Zevnik B, Stephan H, Mannherz HG, Moroy T (2000) Features of systemic lupus erythematosus in Dnase1-deficient mice. Nat Genet 25(2):177–181

Newton K (2015) RIPK1 and RIPK3: critical regulators of inflammation and cell death. Trends Cell Biol 25(6):347–353

Nials AT, Uddin S (2008) Mouse models of allergic asthma: acute and chronic allergen challenge. Dis Model Mech 1(4–5):213–220

Nikoletopoulou V, Markaki M, Palikaras K, Tavernarakis N (2013) Crosstalk between apoptosis, necrosis and autophagy. Biochim Biophys Acta 1833(12):3448–3459

Noelia A-G, Bensinger SJ, Hong C, Beceiro S, Bradley MN, Zelcer N, Deniz J, Ramirez C, Diaz M, Gallardo G, de Galarreta CR, Salazar J, Lopez F, Edwards P, Parks J, Andujar M, Tontonoz P, Castrillo A (2009) Apoptotic cells promote their own clearance and immune tolerance through activation of the nuclear receptor LXR.. Immunity 31(2):245–258

Nordmann M, Cabrera M, Perz A, Brocker C, Ostrowicz C, Engelbrecht-Vandre S, Ungermann C (2010) The Mon1-Ccz1 complex is the GEF of the late endosomal Rab7 homolog Ypt7. Curr Biol 20(18):1654–1659

Obeid M, Tesniere A, Ghiringhelli F, Fimia GM, Apetoh L, Perfettini JL, Castedo M, Mignot G, Panaretakis T, Casares N, Metivier D, Larochette N, van Endert P, Ciccosanti F, Piacentini M, Zitvogel L, Kroemer G (2007) Calreticulin exposure dictates the immunogenicity of cancer cell death. Nat Med 13(1):54–61

Oberst A, Dillon CP, Weinlich R, McCormick LL, Fitzgerald P, Pop C, Hakem R, Salvesen GS, Green DR (2011) Catalytic activity of the caspase-8-FLIP(L) complex inhibits RIPK3-dependent necrosis. Nature 471(7338):363–367

O'Brien BA, Geng X, Orteu CH, Huang Y, Ghoreishi M, Zhang Y, Bush JA, Li G, Finegood DT, Dutz JP (2006) A deficiency in the in vivo clearance of apoptotic cells is a feature of the NOD mouse. J Autoimmun 26(2):104–115

Olazabal IM, Caron E, May RC, Schilling K, Knecht DA, Machesky LM (2002) Rho-kinase and myosin-II control phagocytic cup formation during CR, but not FcgammaR, phagocytosis. Curr Biol 12(16):1413–1418

Oldenborg PA, Zheleznyak A, Fang YF, Lagenaur CF, Gresham HD, Lindberg FP (2000) Role of CD47 as a marker of self on red blood cells. Science 288(5473):2051–2054

Oram JF, Heinecke JW (2005) ATP-binding cassette transporter A1: a cell cholesterol exporter that protects against cardiovascular disease. Physiol Rev 85(4):1343–1372

Park D, Tosello-Trampont AC, Elliott MR, Lu M, Haney LB, Ma Z, Klibanov AL, Mandell JW, Ravichandran KS (2007) BAI1 is an engulfment receptor for apoptotic cells upstream of the ELMO/Dock180/Rac module. Nature 450(7168):430–434

Park SY, Jung MY, Kim HJ, Lee SJ, Kim SY, Lee BH, Kwon TH, Park RW, Kim IS (2008a) Rapid cell corpse clearance by stabilin-2, a membrane phosphatidylserine receptor. Cell Death Differ 15(1):192–201

Park SY, Kang KB, Thapa N, Kim SY, Lee SJ, Kim IS (2008b) Requirement of adaptor protein GULP during stabilin-2-mediated cell corpse engulfment. J Biol Chem 283(16):10593–10600

Park D, Hochreiter-Hufford A, Ravichandran KS (2009) The phosphatidylserine receptor TIM-4 does not mediate direct signaling. Curr Biol 19(4):346–351

Park MC, Kwon YJ, Chung SJ, Park YB, Lee SK (2010a) Liver X receptor agonist prevents the evolution of collagen-induced arthritis in mice. Rheumatology (Oxford) 49(5):882–890

Park SY, Kim SY, Kang KB, Kim IS (2010b) Adaptor protein GULP is involved in stabilin-1-mediated phagocytosis. Biochem Biophys Res Commun 398(3):467–472

Parnaik R, Raff MC, Scholes J (2000) Differences between the clearance of apoptotic cells by professional and non-professional phagocytes. Curr Biol 10(14):857–860

Peter C, Waibel M, Radu CG, Yang LV, Witte ON, Schulze-Osthoff K, Wesselborg S, Lauber K (2008) Migration to apoptotic "find-me" signals is mediated via the phagocyte receptor G2A. J Biol Chem 283(9):5296–5305

Peter C, Wesselborg S, Herrmann M, Lauber K (2010) Dangerous attraction: phagocyte recruitment and danger signals of apoptotic and necrotic cells. Apoptosis 15(9):1007–1028

Poon IK, Lucas CD, Rossi AG, Ravichandran KS (2014) Apoptotic cell clearance: basic biology and therapeutic potential. Nat Rev Immunol 14(3):166–180

Pujol-Autonell I, Ampudia RM, Planas R, Marin-Gallen S, Carrascal J, Sanchez A, Marin A, Puig-Domingo M, Pujol-Borrell R, Verdaguer J, Vives-Pi M (2013) Efferocytosis promotes suppressive effects on dendritic cells through prostaglandin E2 production in the context of autoimmunity. PLoS ONE 8(5):e63296

Rattray M, Cohen GM (1997) Apoptosis and necrosis in toxicology: a continuum or distinct modes of cell death? Pharmacol Ther 75(3):153–177

Ravichandran KS (2010) Find-me and eat-me signals in apoptotic cell clearance: progress and conundrums. J Exp Med 207(9):1807–1817

Razani B, Feng C, Coleman T, Emanuel R, Wen H, Hwang S, Ting JP, Virgin HW, Kastan MB, Semenkovich CF (2012) Autophagy links inflammasomes to atherosclerotic progression. Cell Metab 15(4):534–544

Reddien PW, Horvitz HR (2000) CED-2/CrkII and CED-10/Rac control phagocytosis and cell migration in *Caenorhabditis elegans*. Nat Cell Biol 2(3):131–136

Reddien PW, Horvitz HR (2004) The engulfment process of programmed cell death in caenorhabditis elegans. Annu Rev Cell Dev Biol 20:193–221

Riedl SJ, Shi Y (2004) Molecular mechanisms of caspase regulation during apoptosis. Nat Rev Mol Cell Biol 5(11):897–907

Riento K, Ridley AJ (2003) Rocks: multifunctional kinases in cell behaviour. Nat Rev Mol Cell Biol 4(6):446–456

Rodriguez DA, Weinlich R, Brown S, Guy C, Fitzgerald P, Dillon CP, Oberst A, Quarato G, Low J, Cripps JG, Chen T, Green DR (2015) Characterization of RIPK3-mediated phosphorylation of the activation loop of MLKL during necroptosis. Cell Death Differ

Rodriguez-Manzanet R, Sanjuan MA, Wu HY, Quintana FJ, Xiao S, Anderson AC, Weiner HL, Green DR, Kuchroo VK (2010) T and B cell hyperactivity and autoimmunity associated with niche-specific defects in apoptotic body clearance in TIM-4-deficient mice. Proc Natl Acad Sci U S A 107(19):8706–8711

Roszer T, Menendez-Gutierrez MP, Lefterova MI, Alameda D, Nunez V, Lazar MA, Fischer T, Ricote M (2011) Autoimmune kidney disease and impaired engulfment of apoptotic cells in mice with macrophage peroxisome proliferator-activated receptor gamma or retinoid X receptor alpha deficiency. J Immunol 186(1):621–631

Rothlin CV, Ghosh S, Zuniga EI, Oldstone MB, Lemke G (2007) TAM receptors are pleiotropic inhibitors of the innate immune response. Cell 131(6):1124–1136

Rumore PM, Steinman CR (1990) Endogenous circulating DNA in systemic lupus erythematosus. Occurrence as multimeric complexes bound to histone. J Clin Invest 86(1):69–74

Sanjuan MA, Dillon CP, Tait SW, Moshiach S, Dorsey F, Connell S, Komatsu M, Tanaka K, Cleveland JL, Withoff S, Green DR (2007) Toll-like receptor signalling in macrophages links the autophagy pathway to phagocytosis. Nature 450(7173):1253–1257

Santiago C, Ballesteros A, Martinez-Munoz L, Mellado M, Kaplan GG, Freeman GJ, Casasnovas JM (2007) Structures of T cell immunoglobulin mucin protein 4 show a metal-Ion-dependent ligand binding site where phosphatidylserine binds. Immunity 27(6):941–951

Savill J, Dransfield I, Gregory C, Haslett C (2002) A blast from the past: clearance of apoptotic cells regulates immune responses. Nat Rev Immunol 2(12):965–975

Schrijvers DM, De Meyer GR, Kockx MM, Herman AG, Martinet W (2005) Phagocytosis of apoptotic cells by macrophages is impaired in atherosclerosis. Arterioscler Thromb Vasc Biol 25(6):1256–1261

Scott RS, McMahon EJ, Pop SM, Reap EA, Caricchio R, Cohen PL, Earp HS, Matsushima GK (2001) Phagocytosis and clearance of apoptotic cells is mediated by MER. Nature 411 (6834):207–211

Segawa K, Suzuki J, Nagata S (2011) Constitutive exposure of phosphatidylserine on viable cells. Proc Natl Acad Sci U S A 108(48):19246–19251

Segawa K, Kurata S, Yanagihashi Y, Brummelkamp TR, Matsuda F, Nagata S (2014) Caspase-mediated cleavage of phospholipid flippase for apoptotic phosphatidylserine exposure. Science 344(6188):1164–1168

Shiozaki EN, Chai J, Shi Y (2002) Oligomerization and activation of caspase-9, induced by Apaf-1 CARD. Proc Natl Acad Sci U S A 99(7):4197–4202

Simpson JL, Gibson PG, Yang IA, Upham J, James A, Reynolds PN, Hodge S, A. S. R. Group (2013) Impaired macrophage phagocytosis in non-eosinophilic asthma. Clin Exp Allergy 43(1):29–35

Sokolowski JD, Mandell JW (2011) Phagocytic clearance in neurodegeneration. Am J Pathol 178(4):1416–1428

Su HP, Nakada-Tsukui K, Tosello-Trampont AC, Li Y, Bu G, Henson PM, Ravichandran KS (2002) Interaction of CED-6/GULP, an adapter protein involved in engulfment of apoptotic cells with CED-1 and CD91/low density lipoprotein receptor-related protein (LRP). J Biol Chem 277(14):11772–11779

Sun Q, Westphal W, Wong KN, Tan I, Zhong Q (2010) Rubicon controls endosome maturation as a Rab7 effector. Proc Natl Acad Sci U S A 107(45):19338–19343

Suzuki J, Umeda M, Sims PJ, Nagata S (2010) Calcium-dependent phospholipid scrambling by TMEM16F. Nature 468(7325):834–838

Suzuki J, Denning DP, Imanishi E, Horvitz HR, Nagata S (2013) Xk-related protein 8 and CED-8 promote phosphatidylserine exposure in apoptotic cells. Science 341(6144):403–406

Tabas I (2005) Consequences and therapeutic implications of macrophage apoptosis in atherosclerosis: the importance of lesion stage and phagocytic efficiency. Arterioscler Thromb Vasc Biol 25(11):2255–2264

Tait SW, Green DR (2010) Mitochondria and cell death: outer membrane permeabilization and beyond. Nat Rev Mol Cell Biol 11(9):621–632

Tang C, Liu Y, Kessler PS, Vaughan AM, Oram JF (2009) The macrophage cholesterol exporter ABCA1 functions as an anti-inflammatory receptor. J Biol Chem 284(47):32336–32343

Taylor RC, Cullen SP, Martin SJ (2008) Apoptosis: controlled demolition at the cellular level. Nat Rev Mol Cell Biol 9(3):231–241

Thorp E, Cui D, Schrijvers DM, Kuriakose G, Tabas I (2008) Mertk receptor mutation reduces efferocytosis efficiency and promotes apoptotic cell accumulation and plaque necrosis in atherosclerotic lesions of apoe-/- mice. Arterioscler Thromb Vasc Biol 28(8):1421–1428

Truman LA, Ford CA, Pasikowska M, Pound JD, Wilkinson SJ, Dumitriu IE, Melville L, Melrose LA, Ogden CA, Nibbs R, Graham G, Combadiere C, Gregory CD (2008) CX3CL1/fractalkine is released from apoptotic lymphocytes to stimulate macrophage chemotaxis. Blood 112(13):5026–5036

Vago JP, Nogueira CR, Tavares LP, Soriani FM, Lopes F, Russo RC, Pinho V, Teixeira MM, Sousa LP (2012) Annexin A1 modulates natural and glucocorticoid-induced resolution of inflammation by enhancing neutrophil apoptosis. J Leukoc Biol 92(2):249–258

van Bruggen MC, Kramers C, Walgreen B, Elema JD, Kallenberg CG, van den Born J, Smeenk RJ, Assmann KJ, Muller S, Monestier M, Berden JH (1997) Nucleosomes and histones are present in glomerular deposits in human lupus nephritis. Nephrol Dial Transplant 12(1):57–66

van den Eijnde SM, Boshart L, Baehrecke EH, De Zeeuw CI, Reutelingsperger CP, Vermeij-Keers C (1998) Cell surface exposure of phosphatidylserine during apoptosis is phylogenetically conserved. Apoptosis 3(1):9–16

van den Eijnde SM, van den Hoff MJ, Reutelingsperger CP, van Heerde WL, Henfling ME, Vermeij-Keers C, Schutte B, Borgers M, Ramaekers FC (2001) Transient expression of phosphatidylserine at cell-cell contact areas is required for myotube formation. J Cell Sci 114 (Pt 20):3631–3642

Vandivier RW, Fadok VA, Hoffmann PR, Bratton DL, Penvari C, Brown KK, Brain JD, Accurso FJ, Henson PM (2002a) Elastase-mediated phosphatidylserine receptor cleavage impairs apoptotic cell clearance in cystic fibrosis and bronchiectasis. J Clin Invest 109(5):661–670

Vandivier RW, Fadok VA, Ogden CA, Hoffmann PR, Brain JD, Accurso FJ, Fisher JH, Greene KE, Henson PM (2002b) Impaired clearance of apoptotic cells from cystic fibrosis airways. Chest 121(3 Suppl):89S

Varfolomeev EE, Schuchmann M, Luria V, Chiannilkulchai N, Beckmann JS, Mett IL, Rebrikov D, Brodianski VM, Kemper OC, Kollet O, Lapidot T, Soffer D, Sobe T, Avraham KB, Goncharov T, Holtmann H, Lonai P, Wallach D (1998) Targeted disruption of the mouse Caspase 8 gene ablates cell death induction by the TNF receptors, Fas/Apo1, and DR3 and is lethal prenatally. Immunity 9(2):267–276

Venegas V, Zhou Z (2007) Two alternative mechanisms that regulate the presentation of apoptotic cell engulfment signal in *Caenorhabditis elegans*. Mol Biol Cell 18(8):3180–3192

Vercammen D, Beyaert R, Denecker G, Goossens V, Van Loo G, Declercq W, Grooten J, Fiers W, Vandenabeele P (1998) Inhibition of caspases increases the sensitivity of L929 cells to necrosis mediated by tumor necrosis factor. J Exp Med 187(9):1477–1485

Vives-Pi M, Rodriguez-Fernandez S, Pujol-Autonell I (2015) How apoptotic beta-cells direct immune response to tolerance or to autoimmune diabetes: a review. Apoptosis 20(3):263–272

Vucic E, Calcagno C, Dickson SD, Rudd JH, Hayashi K, Bucerius J, Moshier E, Mounessa JS, Roytman M, Moon MJ, Lin J, Ramachandran S, Tanimoto T, Brown K, Kotsuma M, Tsimikas S, Fisher EA, Nicolay K, Fuster V, Fayad ZA (2012) Regression of inflammation in atherosclerosis by the LXR agonist R211945: a noninvasive assessment and comparison with atorvastatin. JACC Cardiovasc Imaging 5(8):819–828

Wang H, Sun L, Su L, Rizo J, Liu L, Wang LF, Wang FS, Wang X (2014) Mixed lineage kinase domain-like protein MLKL causes necrotic membrane disruption upon phosphorylation by RIP3. Mol Cell 54(1):133–146

Weinlich R, Green DR (2014) The two faces of receptor interacting protein kinase-1. Mol Cell 56(4):469–480

Weinlich R, Oberst A, Dillon CP, Janke LJ, Milasta S, Lukens JR, Rodriguez DA, Gurung P, Savage C, Kanneganti TD, Green DR (2013) Protective roles for caspase-8 and cFLIP in adult homeostasis. Cell Rep 5(2):340–348

Witting A, Muller P, Herrmann A, Kettenmann H, Nolte C (2000) Phagocytic clearance of apoptotic neurons by microglia/brain macrophages in vitro: involvement of lectin-, integrin-, and phosphatidylserine-mediated recognition. J Neurochem 75(3):1060–1070

Wu YC, Tsai MC, Cheng LC, Chou CJ, Weng NY (2001) C. elegans CED-12 acts in the conserved crkII/DOCK180/Rac pathway to control cell migration and cell corpse engulfment. Dev Cell 1(4):491–502

Wu Y, Singh S, Georgescu MM, Birge RB (2005) A role for Mer tyrosine kinase in alphavbeta5 integrin-mediated phagocytosis of apoptotic cells. J Cell Sci 118(Pt 3):539–553

Wu J, Huang Z, Ren J, Zhang Z, He P, Li Y, Ma J, Chen W, Zhang Y, Zhou X, Yang Z, Wu SQ, Chen L, Han J (2013) Mlkl knockout mice demonstrate the indispensable role of Mlkl in necroptosis. Cell Res 23(8):994–1006

Yang CS, Lee JS, Rodgers M, Min CK, Lee JY, Kim HJ, Lee KH, Kim CJ, Oh B, Zandi E, Yue Z, Kramnik I, Liang C, Jung JU (2012) Autophagy protein Rubicon mediates phagocytic NADPH oxidase activation in response to microbial infection or TLR stimulation. Cell Host Microbe 11 (3):264–276

Zhang SQ, Kovalenko A, Cantarella G, Wallach D (2000) Recruitment of the IKK signalosome to the p55 TNF receptor: RIP and A20 bind to NEMO (IKKgamma) upon receptor stimulation. Immunity 12(3):301–311

Zhong Y, Wang QJ, Li X, Yan Y, Backer JM, Chait BT, Heintz N, Yue Z (2009) Distinct regulation of autophagic activity by Atg14L and Rubicon associated with Beclin 1-phosphatidylinositol-3-kinase complex. Nat Cell Biol 11(4):468–476

Zhou XJ, Lu XL, Lv JC, Yang HZ, Qin LX, Zhao MH, Su Y, Li ZG, Zhang H (2011) Genetic association of PRDM1-ATG5 intergenic region and autophagy with systemic lupus erythematosus in a Chinese population. Ann Rheum Dis 70(7):1330–1337

Zhu X, Owen JS, Wilson MD, Li H, Griffiths GL, Thomas MJ, Hiltbold EM, Fessler MB, Parks JS (2010) Macrophage ABCA1 reduces MyD88-dependent Toll-like receptor trafficking to lipid rafts by reduction of lipid raft cholesterol. J Lipid Res 51(11):3196–3206

Zizzo G, Hilliard BA, Monestier M, Cohen PL (2012) Efficient clearance of early apoptotic cells by human macrophages requires M2c polarization and MerTK induction. J Immunol 189(7):3508–3520

Emerging Roles for RIPK1 and RIPK3 in Pathogen-Induced Cell Death and Host Immunity

Danish Saleh and Alexei Degterev

Abstract Receptor-interacting protein kinases 1 and 3 (RIPK1 and RIPK3) are homologous serine–threonine kinases that were recognized for their roles in directing programmed necrotic cell death or necroptosis under a broad range of pathologic settings. Emerging evidence suggests new physiologic roles for RIPK1 and RIPK3 in mediating cell death of innate immune responses. Our review discusses current evidence on the mechanisms and the impact of RIPK1- and/or RIPK3-dependent cell death in responses to a variety of viral and bacterial pathogens. Furthermore, the discussion also summarizes emerging roles for RIPK1 and RIPK3 in other facets of host immunity, including the maintenance of epithelial barrier function and pro-inflammatory processes that may, in some cases, manifest independent of cell death. Finally, we briefly consider the therapeutic opportunities in targeting RIPK1- and RIPK3-dependent processes in infection and immunity.

Contents

D. Saleh
Medical Scientist Training Program, Sackler School of Graduate Biomedical Sciences, Tufts University School of Medicine, Boston, MA 02111, USA

D. Saleh
Program in Neuroscience, Sackler School of Graduate Biomedical Sciences, Tufts University, Boston, MA 02111, USA

A. Degterev (✉)
Department of Developmental, Molecular and Chemical Biology, Tufts University School of Medicine, Boston, MA 02111, USA
e-mail: alexei.degterev@tufts.edu

Current Topics in Microbiology and Immunology (2017) 403:37–76
DOI 10.1007/82_2015_449

Published Online: 15 September 2015

1 Introduction

Receptor-interacting protein kinase 1 and 3 (RIPK1 and RIPK3) are homologous serine–threonine kinases belonging to the RIPK family of enzymes. These kinases garnered initial interest as key regulators of death-receptor-induced programmed necrosis pathway, termed "necroptosis" (Vanden Berghe et al. 2014; Christofferson et al. 2010). Because necrotic cell death plays an important role in human disease, pathologic contributions of RIPK1 and RIPK3 have attracted major interest. Experimental approaches utilizing genetic deletion of RIPK3, expression of kinase-dead RIPK1, or use of small molecule RIPK1 inhibitors, necrostatins, in models of human disease, indeed suggested important contributions of these kinases in many pathologic states, including ischemia–reperfusion injuries, atherosclerosis, pancreatitis, multiple sclerosis, inflammatory bowel diseases, and others (Linkermann and Green 2014; You et al. 2008; Degterev et al. 2005; Ofengeim et al. 2015). While physiologic contexts involving RIPK1/3 regulation remain less clear, developing evidence, summarized in this review, suggests that these molecules may emerge as important players in regulation of the pathogen response and host immunity.

RIPK1 is comprised of three functionally distinct domains: an N-terminal kinase domain, an intermediate domain, and a C-terminal death domain. Kinase function of RIPK1 is important for tumor necrosis factor receptor (TNFR)-dependent necroptosis as well as apoptosis (Christofferson et al. 2014; Ofengeim and Yuan 2013). The serine/threonine kinase domain includes an aspartate–leucine–glycine (DLG) motif-containing binding pocket for the allosteric RIPK1 kinase inhibitor, Necrostatin-1 (Nec-1) (Degterev et al. 2008, 2013; Xie et al. 2013a, b). Nec-1 has been widely used to define kinase-dependent functions of RIPK1; however, the recent generation of RIPK1 kinase-inactive mouse models has expanded the repertoire of tools available to distinguish kinase dependent from kinase-independent functions of this protein (Polykratis et al. 2014; Berger et al. 2014). The intermediate domain of RIPK1 contains a RIP-homotypic interaction motif (RHIM), which facilitates RIPK1 interaction with other RHIM-domain-containing proteins, including DAI, TRIF, and RIPK3. RHIM-domain interactions are required for "amyloid-like" RIPK1/RIPK3 necrosome complex formation and RIPK1- and/or RIPK3-dependent death signaling (Kaiser et al. 2013; Li et al. 2012; Wu et al. 2014; Upton et al. 2012). Lastly, the C-terminal death domain mediates interaction of RIPK1 with death domain containing receptors such as Fas and TNFR1 (Christofferson et al. 2014; Ofengeim and Yuan 2013). The distinct properties associated with the various domains of RIPK1 enable the enzyme to act as a dynamic regulator of cell death signaling.

RIPK3 shares many similarities to RIPK1, including highly homologous kinase and RHIM-domains. Early work suggested that kinase activities of both proteins are required to initiate necroptosis through phosphorylation-driven assembly of a necrosome complex (Xie et al. 2013a, b; Cho et al. 2010). However, more recently, distinct roles for RIPK3 have been identified setting it apart from RIPK1 functionally. For example, RIPK3 lacks the death domain present in RIPK1. Additionally, the field now appreciates that execution of TNF-α-induced necroptosis requires RIPK3-dependent phosphorylation of mixed-lineage kinase domain-like (MLKL), a downstream effector of necroptosis, with RIPK1 playing a more upstream role in initiating RIPK3 signaling (Xie et al. 2013a, b; Wu et al. 2013). MLKL is a requisite executioner of necroptosis: Phosphorylation-dependent oligomerization and translocation of MLKL to the plasma membrane are directly linked to increase in plasma membrane permeability and reactive oxygen species generation (Hildebrand et al. 2014; Li et al. 2014; Murphy et al. 2013; Cai et al. 2014; Zhao et al. 2012). Notably, RIPK3 can also activate necroptosis independent of RIPK1 (Kaiser et al. 2013; Upton et al. 2012; Wu et al. 2014). However, kinase activity of RIPK3 is not required for RIPK1 kinase-dependent apoptosis (Dondelinger et al. 2013).

Cell death and inflammation are critical components of host immune response against invading microbial pathogens. Programmed death of infected cells releases intracellular microbes for clearance by immune cells and restricts microbial proliferation (Lamkanfi and Dixit 2010; Sridharan and Upton 2014). Pathogen-associated inflammatory changes modify host cell programming and activate cell-mediated immunity, both of which promote systemic antimicrobial states. Localized inflammatory responses also promote tissue repair. In fact, pro-inflammatory necroptotic cell death uniquely induces inflammation and promotes tissue repair concurrent with cellular demise (Kaczmarek et al. 2013; Moriwaki et al. 2014).

Epithelial barrier integrity is also a component of the innate immune system (Kumar et al. 2010; Nestle et al. 2009; Elias2007; Macdonald and Monteleone 2005; Kaser et al. 2010). Appropriate barrier function serves to exclude microbial pathogens from entering and threatening the host. In circumstances of epithelial injury or dysfunction, the host susceptibility to infection is immediately raised.

The forthcoming review explores emerging roles of RIPK1 and RIPK3 in three facets of host immunity. The chapter begins by discussing mechanisms of RIPK1/3-dependent cell death in limiting pathogen dissemination and affecting host organism outcomes. Moreover, the discussion explores other roles for RIPK1 and

RIPK3 in host immunity, including epithelial barrier function and inflammation. Specifically, this chapter includes newly identified roles for RIPK1 and RIPK3 in shielding the host against invading pathogens by regulating epithelial barrier function. Furthermore, this text addresses emerging roles of RIPK1 and RIPK3 in directing inflammatory processes that may, in some cases, manifest independent of cell death. In conclusion, the chapter also comments on therapeutic opportunities in targeting RIPK1- and RIPK3-dependent processes during infection.

2 Pathogen-Induced Cell Death

2.1 Viral-Induced Cell Death

The pathogenesis of viral species is uniquely dependent on their ability to replicate within host cells. Accordingly, viruses have evolved mechanisms inhibiting cell death and prolonging survival of host cells to ensure reliable replication (Galluzzi et al. 2010; Roy and Mocarski 2007; Lamkanfi and Dixit 2010). Viruses equipped to inhibit apoptosis are commonplace and are described extensively in the literature (Mocarski et al. 2012; Lamkanfi and Dixit 2010). Among known examples, Kaposi's sarcoma herpesvirus (HSV-8) and human poxvirus molluscum contagiosum encode viral FLICE-inhibitory protein (v-FLIP) family members; these molecules act in a dominant-negative fashion to inhibit caspase-8 and caspase-10 recruitment to the death-inducing signaling complex (DISC) (Bertin et al. 1997; Hu et al. 1997; Thome et al. 1997; Shisler and Moss 2001; Thurau et al. 2006). Orthopoxviruses, including cowpox, vaccinia, ectromelia, and rabbitpox viruses bear homologous serine protease (serpin) inhibitors of caspases (Dobbelstein et al. 1996; Macen et al. 1996; Turner et al. 2000; Best 2008). These peptides are among a broad class of serpin protease inhibitors that bind within the active site of target enzymes rendering them inactive. Herpesvirus family members, HSV-1 and HSV-2, carry a caspase-8 inhibitor gene, UL-39, which encodes ribonucleotide reductase subunit protein, R1, also known as ICP6 or ICP10 for HSV-1 and HSV-2, respectively (Dufour et al. 2011; Mocarski et al. 2012). A third herpesviridae, cytomegalovirus (CMV), encodes a distinct inhibitor of caspase-8 function and viral inhibitor of caspase-8 activation (vICA) (Skaletskaya et al. 2001). Caspase-8 inhibitors are also present in adenoviruses and human papillomavirus-16 (HPV-16), a strain commonly associated with cervical cancer (Kabsch and Alonso 2002). Baculoviruses express inhibitors of apoptosis, including a homologue of human X-linked inhibitor of apoptosis (XIAP) and a pan-caspase inhibitor, p35 (Birnbaum et al. 1994; Deveraux and Reed 1999; Crook et al. 1993). Apoptosis inhibitors have also been identified in the influenza A virus (Yatim and Albert 2011). Although this list is not exhaustive, it serves to illustrate that apoptosis inhibitors are common in viral species.

Emerging evidence suggests that caspase-independent death may become paramount for host defense in the presence of anti-apoptotic signals or caspase inhibitors. For example, fibroblasts infected with vaccinia virus underwent necrotic cell death that was reduced in cells infected with a mutant lacking caspase inhibitor B13R (Li and Beg 2000). In vaccinia infection, RIPK1- and RIPK3-dependent necroptosis is required to contain viral dissemination; RIPK1 kinase-inactive mice of the D138N flavor ($RIPK1^{D138N}$) and RIPK3 knockout ($RIPK3^{-/-}$) mice displayed a 10-fold to 100-fold increase in viral titers compared to wild-type (WT) counterparts (Cho et al. 2010; Polykratis et al. 2014). $RIPK3^{-/-}$ mice were also observed to have diminished survival and increased viral dissemination compared to that observed in $RIPK1^{D138N}$ mice suggesting that RIPK3 may play a more prominent role in virus-induced necroptosis. Similarly, cell death induced by MCMV virus, lacking RIPK inhibitor M45, has been found to involve direct activation of RIPK3 by viral DNA sensor DAI (Upton et al. 2012). These data suggest that RIPK1 and RIPK3 may play significant roles in anti-viral responses, but their activities may not be linearly connected as has been proposed in TNF-α-induced cell death (Vandenabeele et al. 2010a).

2.1.1 HSV

Herpes simplex viruses, HSV-1 and HSV-2, are viral pathogens commonly recognized for causing tissue ulceration around the oral cavity and genitals, respectively (Klein 2015a; Albrecht 2014). More severely, HSV-1 can also manifest as a life-threatening encephalitis following inoculation of the oralpharyngeal cavity. Notably, HSV-1 encephalitis is regarded as the most common cause of sporadic encephalitis in the world (Klein 2015b).

Molecular studies have demonstrated that in murine cells, HSV-1 induces programmed necrosis which limits viral replication (Wang et al. 2014b; Huang et al. 2015). RIPK3 deletion completely protected mouse embryonic fibroblasts (MEFs) and L929 cells from HSV-1-induced cell death. This function of RIPK3 is attributed to its kinase activity as cells expressing a kinase-dead mutant were just as viable as $RIPK3^{-/-}$ counterparts. L929 cells lacking RIPK1 were partially protected from cell death indicating that RIPK1 is also important for HSV-1-induced cell death. Not surprisingly, HSV-1-induced cell death also required canonical effector of necroptosis, MLKL. Knockdown of MLKL in WT MEFs or L929 cells protected against HSV-1-induced cell death. Importantly, HSV-1-induced cell death effectively limited viral propagation as $RIPK3^{-/-}$ L929 cells generated larger viral titers than WT counterparts post-infection (Huang et al. 2015; Wang et al. 2014b).

Mechanistic analyses revealed that HSV-1-induced cell death was dependent on RHIM-domain interactions. MEFs expressing RHIM-domain mutant of RIPK3 were protected from HSV-1 infection-induced cell death. Authors identified that the viral RHIM-domain containing protein, ribonucleotide reductase subunit 1 (R1), also known as ICP6, induced necrosis in target cells (Wang et al. 2014b; Huang et al. 2015). Furthermore, HSV-1 carrying a RHIM-domain mutant of ICP6

(ICP6mutRHIM) did not cause cell death. RIPK1 and RIPK3 were co-immunoprecipitated with ICP6 in HSV-1 infected MEFs; however, no interaction was detected upon infection with virus carrying ICP6mutRHIM (Huang et al. 2015). Huang and colleagues further demonstrated that ICP6 can induce RIPK1–RIPK3 heterointeractions and RIPK3–RIPK3 homointeractions in HEK293T cells. These findings suggest that HSV-1-induced programmed necrosis requires RHIM-domain interactions of ICP6 with RIPK1 and RIPK3, which leads to induction of necroptosis in infected cells and limits viral infections (Table 1). This presents an interesting mechanism linked to the life cycle of HSVs as expression of ICP6 has been found to be essential for the reactivation of the virus in quiescent cells and has also been reported to inhibit caspase-8 activation (Goldstein and Weller 1988; Langelier et al. 2002; Dufour et al. 2011).

Necroptosis improved host health and survival while limiting viral titers and propagation in vivo. RIPK3$^{-/-}$ animals lost more body weight following infection compared to WT counterparts. Moreover, HSV-1 viral DNA was detected in greater abundance in serum, liver, and nervous tissue in RIPK3$^{-/-}$ animals suggesting increased viral propagation. Greater viral dissemination corresponded to worse outcomes as RIPK3$^{-/-}$ animals had diminished survival compared to controls. Concordantly, viral dissemination was increased and animal survival decreased in WT mice infected with ICP6mutRHIM HSV-1 compared to WT HSV-1.

Interestingly, while two groups reported that HSV-1 is able to effectively induce cell death in murine cells, HSV-1-induced cell death was not similarly observed in human cells (Guo et al. 2015; Huang et al. 2015). Data showed that RHIM-domains of ICP6 in HSV-1 and ICP10 in HSV-2 inhibited TNF-α-induced RIPK3-dependent necroptosis in human HT29 cells by competing for RHIM-domain interactions of RIPK1 and RIPK3 (Guo et al. 2011). In suit, ICP6mutRHIM HSV-1 unleashed TNF-α-induced RIPK3 kinase-dependent cell death and limited viral titers (Table 1) (Guo et al. 2015).

In sum, ICP6 induced necroptosis in a RHIM-dependent manner in murine cells while serving the opposite function, by blocking necroptosis, in human cells. These seemingly contradictory observations stress the versatility of the RHIM-domain in necroptosis signaling. These observations also suggest that important, yet currently unknown, mechanistic differences must exist in how human and mouse RHIM-domains interact. Irrespective of further mechanistic clarification, these findings may be reconciled by the fact that HSV-1 is a natural human pathogen and thus may have evolved to evade human and not mouse RIPK1- and RIPK3-dependent necroptosis. Conversely, HSV-1 may have emerged as a human pathogen because of its ability to block necroptosis specifically in human cells. Although differences between the abilities of host–pathogen interactions in humans and mice to induce necroptosis are not yet well-established, these differences may materialize as important clues in defining innate immune roles of RIPK1 and RIPK3.

Table 1 Specific roles of RIPK1 and/or RIPK3 in microbial pathogenesis

Microbe	Host species	Caspase-8 inhibitor	Mechanism of RIPK inhibition	Mechanism of RIPK activation	RIPKs	Function of RIPK-dependent signaling in host cell	Impact of RIPK-dependent signaling on organism
Vaccinia virus	Murine	SPI-2 or B13R (Serpin)		Undefined	RIPK3	Induces host cell death	Reduced viral titers and improved host survival
Reoviruses (T3D strain)	Murine (L929 cells)	Undetermined		Undefined	RIPK1	Induces host cell death	Undefined
Influenza-A	Human	NS1		Death-receptor ligation/FasL-Fas	RIPK1 and RIPK3	Induces tissue injury/death (in the absence of cIAP2)	Diminished survival (in the absence of cIAP2)
HSV-1	Murine	UL-39/ICP6		RHIM-domain interaction of ICP6	RIPK1 and RIPK3	Induces host cell death	Reduced viral titers and improved survival
HSV-1	Human	UL-39/ICP6	ICP6 competing for RHIM-domain interactions between RIPKs		RIPK1 and RIPK3	Inhibits host cell death	Enhanced viral propagation
HSV-2	Human	UL-39/ICP10	ICP10 competing for RHIM-domain interactions between RIPKs		RIPK1 and RIPK3	Inhibits host cell death	Undefined
MCMV	Murine	vICA, Bax, Bak	M45 disrupting RHIM-domain interactions between DAI and RIPK3		RIPK3	Inhibits host cell death	Enhanced viral propagation and diminished organism survival
hCMV	Human	vICA, Bax, Bak	IE1		Undefined	Inhibits host cell death	Enhanced viral propagation and increased cell death

(continued)

Table 1 (continued)

Microbe	Host species	Caspase-8 inhibitor	Mechanism of RIPK inhibition	Mechanism of RIPK activation	RIPKs	Function of RIPK-dependent signaling in host cell	Impact of RIPK-dependent signaling on organism
Yersinia	Murine	Undefined		YopJ	RIPK1	Induces host cell death	Reduced bacterial propagation, tissue damage, and improved survival
Salmonella	Murine	Undefined		Salmonella-induced IFN-I	RIPK1 and RIPK3	Induces host cell death	Undetermined
E. coli (EPEC)	Murine and Humans	Undefined	Death domain interactions of Nleb1		RIPK1	Inhibits host cell death	Undetermined
Clostridium Perferinges	Porcine	Undefined		CPB	RIPK1	Induces host cell death	Undetermined
Staphylococcus Aureus (Methicillin-Resistant, Strain USA300)	Murine	Undefined		Toxins (Undefined)	RIPK1 and RIPK3	Induces host cell death	Increased tissue destruction, decreased tissue macrophages, and increased bacterial titers
Mycobacterium tuberculosis	Zebra fish	Undefined		Infection-induced TNF-α	RIPK1	Induces ROS generation and host cell death	Undetermined

2.1.2 CMV

CMV is highly prevalent in the general population. Although it is benign in healthy individuals, the virus can be lethal in the immunocompromised, such as AIDS patients and transplant recipients (Friel 2014). Due to its prevalence, CMV is the most common cause of congenital infection in newborns. Mothers experiencing primary CMV infection or viral reactivation are at increased risk for vertical transmission to their unborn children. Congenital infection can lead to permanent damage, marked by hearing loss, vision impairment, and or cognitive retardation (Demmler-Harrison 2015).

In 2010, Kaiser and colleagues reported that murine CMV (MCMV) M45 protein, a RHIM-domain-containing viral inhibitor of RIP activation (vIRA), was required for viral dissemination and decreased host survival (Upton et al. 2010). Further investigation resolved that the RHIM-domain of vIRA inhibited RHIM-domain interactions between DNA-dependent activator of interferon regulatory factors (DAI), an intracellular sensor of viral DNA, and RIPK3 that were required for RIPK3-dependent cell death. In accord, mice infected with MCMV carrying a mutation in the RHIM-domain of M45 protein (M45mutRHIM) had improved survival and lower viral titers in lymphoid organs compared to mice infected with WT MCMV. Similarly, deletion of DAI or RIPK3 rescued the pathogenicity of the M45mutRHIM virus, suggesting that RIPK3-dependent necrosis mitigated viral-induced health decline of the organism and minimized viral dissemination (Upton et al. 2010, 2012). These findings are consistent with a model in which RIPK3-dependent necrotic cell death limits viral replication by triggering necroptosis in the infected cells to protect the host against a pathogen fortified to suppress pro-apoptotic signals (Table 1).

Analogous to MCMV, human CMV (hCMV) is able to block necrotic death in target cells. In 2015, Mocarski's group reported that low passage, newborn human foreskin fibroblasts transduced with hRIPK3 and infected with hCMV were protected from necrotic cell death induced by either a combination of TNF-α, smac mimetics (SM), and zVAD.fmk (zV) or M45mutRHIM MCMV infection. Importantly, in the absence of hCMV, both conditions induced death dependent on the kinase activity of hRIPK3 as death was not observed in cells transduced with a kinase-dead mutant of RIPK3 (hRIPK3 K50A). Remarkably, phosphorylated forms of hRIPK3 and MLKL were still observed in hCMV-infected cells, suggesting that hCMV differs from MCMV in that it blocks programmed necrosis downstream of RIPK3 and MLKL activation through a yet-to-be-defined mechanism. An approach utilizing UV light-induced viral mutagenesis revealed that viral regulatory protein, IE1, known to modulate host gene expression and innate immune signaling, was required by hCMV to block programmed necrosis. Expectedly, deletion of IE1 increased sensitivity to hRIPK3-induced cell death and resulted in decreased viral titers in vitro (Omoto et al. 2015).

Human and murine CMVs apply different strategies in blocking RIPK3-dependent cell death. This may reflect evolutionary variations in viral species to accommodate specificities of host cell biology. For example, although

RHIM-domain interactions may be sufficient to block RIPK3-dependent cell death in murine cells, this may not be true for human hosts. Accordingly, hCMV is armed with mechanisms independent of RHIM-domain interactions to block necroptosis. Similarly, RHIM-domain containing proteins that block HSV-induced cell death in mouse cells fail to do so in human hosts. These observations suggest that fundamental differences in RIPK3-dependent cell death pathways between human and mouse systems are not yet fully understood.

2.1.3 Reoviruses

Mammalian orthoreoviruses (Reoviruses) comprise a family of double-stranded RNA viruses known to cause respiratory and enteric infections. Although most reoviruses are not associated with clinical disease, the rotavirus, a species of reovirus, is the most common cause of gastroenteritis, resulting in diarrhea and fever in children (O'Ryan and Matson 2015).

Reoviruses are known to induce caspase-dependent apoptosis in target cells (Clarke et al. 2003; Richardson-burns et al. 2002; Danthi et al. 2013). However, recent studies have suggested that these viruses may also induce necroptosis. For example, the Type-3 Dearing (T3D) strain induced cell death in murine L929 fibrosarcoma cells that could not be blocked by inhibition of caspases, but was blocked by Nec-1. This observation suggests that the T3D strain can induce cell death in a RIPK1 kinase-dependent manner (Berger and Danthi 2013) (Table 1).

Mechanisms responsible for the activation of necroptosis by reoviruses are currently unknown. Additionally, the functional value of RIPK1 kinase-dependent mechanisms in reoviral disease, including control of viral counts in vivo, duration of infection, tissue histopathology, and health outcomes, remains to be established.

2.1.4 Influenza A

Influenza viruses, part of the broader Orthomyxoviridae family, are enveloped negative-strand RNA viruses that include strains A and B. The virus is readily aerosolized and communicable in the human population, causing a self-limited, acute, seasonal respiratory tract infection, identified as *Influenza* or *Flu*. Patients manifest with upper and/or lower respiratory tract illness in conjunction with signs of systemic illness including fever, headache, muscle pain, and weakness. Certain high-risk patient populations, such as the elderly, pregnant, and immunosuppressed, may develop a more complicated illness, marked by pneumonia or infection of the nervous system (Dolin 2015; Yatim and Albert 2011).

Deletion of RIPK1 ubiquitin ligase, cellular inhibitor of apoptosis 2 (cIAP2), promotes Influenza A-induced respiratory tissue necrosis that is dependent on RIPK1 kinase activity and RIPK3. Studies also found that cIAP2 deficient mice had diminished survival without alterations in viral load (Rodrigue-Gervais et al. 2014). Notably, cIAP2 knockout mice treated with Nec-1 or also carrying a deletion of

RIPK3 were protected from pathologic features of tissue necrosis and displayed improved survival. Significantly, loss of RIPK3 in cIAP2 knockout mice did not alter viral titers, indicating that early death in cIAP2 knockout mice was likely a consequence of respiratory failure as opposed to viral burden. These data provide evidence that in the absence of cIAP2, RIPK1- and RIPK3-dependent cell death are not required to limit viral propagation, but rather to promote lethal tissue injury (Rodrigue-Gervais et al. 2014) (Table 1).

Deletion of cIAP2 may induce RIPK1 and RIPK3 hyperactivity that licenses indirect tissue injury during Influenza-A infection. It is well-established that cIAPs are important physiologic inhibitors of RIPK1- and RIPK3-dependent necroptosis (McComb et al. 2012). Accordingly, RIPKs may be inappropriately activated in the absence of cIAP2. Acute activation of RIPK1- and RIPK3-dependent cell death following viral infection suggests that RIPK1 and RIPK3 functions are specifically modified following viral infection, either directly, as a consequence of viral regulation, or indirectly, secondary to local inflammation and death-receptor ligation. Indeed, deletion of death-receptor ligand, FasL, normalized animal survival, suggesting that RIPK1 and RIPK3 are inappropriately activated downstream of FasL in cIAP2 knockout mice (Rodrigue-Gervais et al. 2014). Furthermore, fluorescence-activated cell sorting (FACS) analysis demonstrated that both uninfected and infected respiratory epithelia of cIAP2 knockout mice were more likely to undergo necrotic cell death than their WT counterparts, suggesting that cell death is induced by an exogenously released factor(s) rather than the virus itself (Rodrigue-Gervais et al. 2014). Lastly, authors found that RIPK3$^{-/-}$ mice did not have surplus viral burden, deficits in viral clearance, or excess tissue injury compared to WT counterparts (Rodrigue-Gervais et al. 2014). Collectively, these data suggest that Influenza-A infection, in the absence of cIAP2, promoted RIPK3-dependent cell death and tissue destruction indirectly, as a consequence of death-receptor ligation.

2.2 *Bacterial-Induced Cell Death*

Bacteria are able to replicate independently of the host cell, and thus, bacterial pathogenesis is oftentimes not strictly dependent on host cell survival. Nevertheless, certain bacterial species survive and proliferate intracellularly. For these species, host cell integrity and longevity may be paramount (Pujol and Bliska 2005). Recently, RIPK-dependent cell death has been identified as an important player in bacterial infection. However, it remains to be fully determined whether infection-associated activation of RIPK1 and RIPK3 signaling serves as a feature of the host response or an exploitable weakness in host defense. Examples of RIPK1- and RIPK3-dependent regulation of the host response against specific bacterial pathogens are examined below.

2.2.1 *Yersinia*

Yersinia species are Gram-negative enteric pathogens that include the human pathogenic forms *Yersinia pestis*, *Yersinia pseudotuberculosis*, and *Yersinia entercolitica*. *Pestis* is notorious for its ferocity and was the root of the black plague in medieval Europe (Sexton 2014). *Peudotuberculosis* and *Entercolitica* produce similar symptoms with fever, abdominal pain, and diarrhea (Tauxe 2013). Uniquely, *Pseudotuberculosis* can also manifest with tuberculosis-like symptoms, including tissue necrosis and granulomas in the spleen, liver, and lymph nodes (Viboud and Bliska 2005). Virulence of these bacteria is dependent on the translocation of a series of pathogenicity factors, called *Yersinia* outer proteins (Yops), which are exported from the bacteria into host cells by a bacterial-encoded protein translocation system, the type-III secretion system (TTSS) (Viboud and Bliska 2005; Zhang et al. 2011; LaRock and Cookson 2012).

Phagocytic cells, in particular macrophages and dendritic cells, play important roles in the pathogenesis of *Yersinia* species. During infection, these host immune cells engulf bacteria and facilitate their systemic spread, specifically to mesenteric lymph nodes, spleen, and liver (Viboud and Bliska 2005). Significantly, all three human pathogenic *Yersinia* species have also been shown to replicate within macrophages and dendritic cells (Pujol and Bliska 2005; Bliska 2003; Pujol et al. 2009). In accord, programmed cell death of phagocytic cells in *Yersinia* infection may be an important feature of the host response.

Recent studies have uncovered important roles for RIPK1 in *Yersinia*-induced macrophage cell death. Pharmacologic inhibition of RIPK1 prevented *Yersinia*-induced cell death in macrophages (Weng et al. 2014; Philip et al. 2014). RIPK1-deficient fetal liver-derived macrophages were also resistant to *Y. pestis*-induced cell death. Notably, *Yersinia*-induced cell death was also shown to require caspase-8 and kinase activity of RIPK3 (Weng et al. 2014). Although macrophages lacking caspase-8 or RIPK3 or treated with RIPK3 inhibitor, GSK'872, were not protected from *Pestis*-induced death, combined loss of caspase-8 and deletion or inhibition of RIPK3 resulted in complete protection (Weng et al. 2014). The sufficiency of RIPK1 kinase and either caspase-8 or RIPK3 for cell death suggests that RIPK1 may function to activate either caspase-8-dependent apoptosis or RIPK3-dependent necroptosis in response to *Yersinia* infection.

Caspase-8 and RIPK3 double-knockout animals had increased susceptibility to infection with *Yersinia* (Weng et al. 2014; Philip et al. 2014). These mice had increased bacterial colony-forming units in spleens, increased inflammatory infiltrates in hepatic tissue sections, and diminished survival. As noted during viral infection, inhibition of cell death during *Yersinia* infection was detrimental to organism survival. Accordingly, elimination of infected host cells is likely important for the control of *Yersinia* as it is for viruses because both pathogens survive and replicate intracellularly. Altogether, these observations suggest that RIPK1-dependent apoptosis requiring caspase-8 or RIPK1-dependent necroptosis requiring RIPK3 may be utilized by the host to combat *Yersinia* infection (Table 1).

Caspase-1-dependent pro-inflammatory cell death, pyroptosis, may also play a role in *Yersinia*-induced, RIPK1-dependent, cell death of phagocytic cells. Experiments showed that in the absence of RIPK1, *Yersinia*-infection resulted in reduced production of the cleaved or active form of caspase-1 (Weng et al. 2014). Although this finding suggests a role for RIPK1 in *Yersinia*-induced caspase-1 activation, it remains to be determined whether pyroptosis actually plays a role in *Yersinia* infection.

Investigators attributed *Yersinia*-induced, RIPK1-dependent cell death to YopJ, one of the Yop pathogenicity factors. Macrophages infected with *Yersinia* mutants lacking functional YopJ (ΔYopJ) were protected from RIPK1-dependent cell death. YopJ is well known for inhibiting pro-inflammatory and pro-survival NFκB and MAPK signaling in target cells. Not surprisingly, this effector has previously also been implicated in inflammasome activation and pyroptotic cell death of host cells (Weng et al. 2014; Philip et al. 2014). Accordingly, host immune cells may have evolved new modalities of YopJ-induced cell death to mitigate bacterial pathogenicity in the face of a bacterial agent, YopJ, which limits inflammatory responses (Table 1).

Regarded together, these observations cement a role for RIPK1-dependent cell death in *Yersinia* infection. Although it is unclear whether RIPK1 induces primarily an apoptosis, necroptosis, or pyroptosis, or perhaps some combination of the three, the *Yersinia* infection model uncovers novel infection-associated cell death regulation by RIPK1. Moreover, appreciating that *Yersinia* species have the capacity to replicate within phagocytic cells, it is not surprising that programmed elimination of infected immune cells protects against dissemination of *Yersinia* in the host.

2.2.2 *Salmonella*

Salmonella enterica species (*Salmonella*) are Gram-negative facultative intracellular bacilli which cause gastroenteritis in humans with symptoms including fever, diarrhea, and abdominal pains (Kotton and Hohmann 2013). Similar to other bacteria which can survive and replicate in host immune cells, *Salmonella* is able to commandeer anti-inflammatory M2 macrophages, spread systemically, and seed lymphoid tissue (Behnsen et al. 2015; Alpuche-Aranda et al. 1994; Nix et al. 2007; Kotton and Hohmann 2013; McCoy et al. 2012). Dissemination of *Salmonella* can promote tissue-specific damage including abscess formation and osteomyelitis as well as systemic infection or sepsis (Hohmann 2014).

A recent study demonstrated that *Salmonella*-induced type I interferon (IFN-I) signaling resulted in macrophage necroptosis (Robinson et al. 2012). Macrophages deficient in the interferon alpha receptor, which is required for sensing IFN-Is, or cultured with anti-IFN-I antibodies, were resistant to *Salmonella*-induced cell death. Inhibition of RIPK1 blocked over 60 % of cell death induced by *Salmonella* in WT macrophages which was similar to the extent of death observed in IFNaR knockout (IFNaR$^{-/-}$) macrophages. Moreover, deletion or knockdown of RIPK3 in macrophages abrogated *Salmonella*-induced cell death (Robinson et al. 2012).

Further investigation uncovered that both IFNaR$^{-/-}$ and RIPK3$^{-/-}$ mice had decreased bacterial load and increased numbers of circulating macrophages compared to WT counterparts. This protective phenotype was likely conferred by the hematopoietic compartment as $INFaR^{-/-}$ bone marrow transplanted into WT mice limited bacterial titers. Curiously, $RIPK3^{-/-}$ mice did not have improved survival indicating that bacterial titers do not necessarily correlate with organism survival in the case of *Salmonella* infection (Robinson et al. 2012).

These paradoxical observations merit investigation into the relationship between *Salmonella* titers and manifestation of disease. One may inquire whether increased macrophage survival in the absence RIPK3 is underestimating total bacterial loads by excluding intracellular bacteria. In this case, necroptosis of anti-inflammatory M2 macrophages may serve to limit intracellular proliferation of *Salmonella* and contain infection. Nevertheless, these data strongly suggest that *Salmonella* infection presents another model of RIPK-dependent innate immune regulation. Interestingly, this regulation manifests by a mechanism that is distinct, and perhaps functionally opposite, from cell death induced by *Yersinia* (Table 1).

2.2.3 *Escherichia Coli*

Escherichia coli (*E. coli*) is a Gram-negative bacterial species that includes a diversity of strains or serotypes. Although the majority of these serotypes are benign, some have procured the capacity to cause serious infection and tissue damage. One such pathogenic strain, *Enteropathogenic E. coli* (*EPEC*), is associated with sporadic diarrheal illness, particularly in young children. The pathogenesis of virulent strains of *E. coli* is attributed to the acquisition of toxic effectors that are injected into target or host cells (Wanke 2013).

Nleb1 is a pathogenicity effector protein found among species of *E. coli* and *Salmonella*. Mechanistic analyses revealed Nleb1 possesses N-acetylglucosamine transferase activity that modifies a conserved arginine residue in death domain containing proteins, including TNF-α receptor (TNFR), TNF receptor-associated death domain (TRADD), FAS receptor (FASR), and RIPK1 (Li et al. 2013; Pearson et al. 2013). Nleb1 enzymatic activity blocked TRADD oligomerization and death-receptor-induced signaling complex (DISC) formation (Li et al. 2013; Pearson et al. 2013). Pearson and colleagues found that Nleb1 co-immunoprecipitated with death domain proteins, TRADD, FADD, and RIPK1, in 293T cells. In addition, Nleb1 abolished death-receptor-induced apoptosis in HeLa cells and TNF-α/zVAD/SMAC mimetic-induced necroptosis in RIPK3-expressing HeLa cells (Li et al. 2013). In murine infection models, *EPEC* either lacking Nleb1 or carrying enzymatically inactive Nleb1 mutants were unable to effectively colonize host gastrointestinal tracts (Li et al. 2013). These findings suggest that multiple cell death pathways, including RIPK3-dependent cell death, may be manipulated by Nleb1 expressing *E. coli* to establish infection, potentiate survival of infected cells, and minimize associated tissue destruction. Specific roles for RIPK1 and RIPK3 in Nleb1-associated infection remain to be elucidated (Table 1).

2.2.4 Pore-Forming Toxin-Producing Bacteria

Pore-forming proteins (PFPs) are bacterial toxins that increase host cell membrane permeability and thereby contribute to bacterial virulence. Some commonly recognizable virulent bacterial species and their significant clinical presentations include the following: *E. coli* (urinary tract infections and gastroenteritis), *Corynebacterium diphtheria* (diphtheria; upper respiratory infections), *Clostridium perferinges* (tissue necrosis and gas gangrene), *Clostridium septicum* (tissue necrosis and gas gangrene), *Bacillus anthracis* (anthrax; pneumonia, gastroenteritis, and/or cutaneous ulcers), *Listeria monocytogenes* (gastroenteritis and meningitis), *Staphylococcus aureus* (cellulitis, impetigo, abscess, and respiratory tract infections), *Vibrio cholera* (diarrhea and dehydration), and *Yersinia pseudotuberculosis* (gastroenteritis) (Bischofberger et al. 2012).

PFP-associated changes in membrane permeability may serve in any of a variety of functions in bacterial pathogenesis. The literature includes precedents of PFPs that direct translocation of effector proteins or other toxins into host cells, facilitate microbial invasion of the intracellular space, disrupt host cell ion homeostasis and energy balance, alter membrane compartment dynamics, and/or promote direct host cell destruction. For example, type-III secretion systems (TTSS) of Gram-negative bacteria, such as *Yersinia* and *Salmonella* species, use PFPs to translocate effector proteins into target cells (Zhang et al. 2011). Alternatively, *L. monocytogenes*' PFP, listeriolysin O (LLO) blocks acquisition of anti-bacterial proteins by *Listeria* containing host cell vacuoles (Bischofberger et al. 2012).

At high PFP concentrations, host cell death occurs by necrosis; however, at low PFP concentrations, cell death has been linked to apoptosis, necroptosis, as well as pyroptosis (Knapp et al. 2010; Kennedy et al. 2009; Lin et al. 2010; Boyden and Dietrich 2006; Craven et al. 2009). In the latter instance, the lack of uniformity in cell death suggests that host cell demise may be part of a regulated host response as opposed to a generalized manifestation of death secondary to excessive cell stress or membrane perforation. Notably, Nec-1 inhibited necrosis induced by *Clostridium perferingees*' PFP, also known as *C. perferingees* β-toxin (CPB), thereby specifying RIPK1 kinase-dependent regulation of CPB-induced necrosis (Autheman et al. 2013).

A study published by Kitur and colleagues suggested that RIPK1- and RIPK3-dependent necrosis may not be beneficial to the host. Kitur observed that toxin-induced programmed necrosis is a major mechanism of lung damage by methicillin-resistant *S. Aureus* (*MRSA or SA*), strain USA300. Authors reported that *SA* infection induced programmed necrosis in THP-1 human monocytes that was inhibited by Nec-1 or the MLKL inhibitor necrosulfonamide (NSF). Cytotoxicity was attributed to pore-forming toxin, α-hemolysin (Hla), and other leukotoxins as macrophage cell death could be induced by bacterial-conditioned media alone, but was abrogated by conditioned media from selective toxin-null mutant strains. Remarkably, murine lung infection models uncovered that RIK3$^{-/-}$ or blockade of necrosis by Nec-1 enhanced bacterial clearance and limited tissue injury. Furthermore, WT mice had diminished macrophage populations in lung tissue compared to $RIPK3^{-/-}$ counterparts following infection (Kitur et al. 2015). These findings suggest that necrotic cell death induced by toxinproducing bacterial species, such as *SA*, may aggravate tissue damage and cause localized immunodeficiency to enhance bacterial infection and thereby compromise host survival (Table 1).

Although the collected data indicate that RIPK1 and RIPK3 may be important in models of PFP-bacterial infection models, it is unclear as to whether RIPK1- and/or RIPK3-dependent cell death generally serves as an advantageous or injurious feature of the host response. The majority of reports available do not explore the contributions of RIPK1- and/or RIPK3-dependent cell death in PFP-associated bacterial dissemination and organism survival. Accordingly, the roles of RIPK1- and/or RIPK3- dependent cell death in PFP-associated bacterial pathogenesis remain unclear.

2.2.5 *Mycobacterium Tuberculosis*

Mycobacterium tuberculosis (*M.tb*) is the causative agent of *Tuberculosis* (*TB*), an infectious disease that is communicable by respiratory secretions in humans. Although *TB* commonly manifests in the lungs because aerosolized particles descend into pulmonary recesses, the infection can disseminate systemically if poorly contained. *TB* is established in one of three forms: primary, latent, and reactivation. Primary *TB* is active or fulminant *TB* that occurs following a new or initial exposure. Latent *TB* is established when a new infection enters a state of dormancy and is thought to occur when invading *M.tb* is confronted by a robust immune response. In this case, the host is neither contagious nor exhibits the symptoms of *TB* and is labeled a carrier. The third form, reactivation *TB*, is defined as the emergence of active or symptomatic *TB* in a carrier (Pozniak 2015; Schluger and Rom).

The genesis of *TB* is a complex process that is poorly understood; however, it is accepted that macrophages are required to contain infection (Ulrichs and Kaufmann 2006). Additionally, TNF-α is recognized as a critical cytokine in the development of *TB*; for example, patients with rheumatoid arthritis on an anti-TNF-α regimen are more likely to develop reactivation of *TB* (Miller and Ernst 2009). Patients with compromised immunity, who are immunodeficient or immunosuppressed, are also at increased risk of developing primary or reactivation *TB* (Pozniak 2015). Moreover, zebra fish models of *Mycobacterium* infection have demonstrated that exogenous TNF-α can limit mycobacterial titers in animals with deficiencies in TNF-α synthesis (Tobin et al. 2010, 2012; Roca and Ramakrishnan 2013). Roca and Ramakrishnan also report that TNF-α induces production of reactive oxygen species (ROS) and thereby augments macrophage bacteriocidal activity and promotes RIPK1- and RIPK3-dependent necroptosis. As TNF-α is the best known inducer of necroptosis in vitro and in vivo, these data may suggest involvement of RIPK1 and/or RIPK3, but this has not yet been established directly. Moreover, associations noted above prompt questions examining the impact of these RIPKs on organism health and survival in *M.tb* infection (Table 1).

2.3 *Summary*

Multiple lines of evidence suggest that activation of RIPK1- and/or RIPK3-dependent cell death is an important consequence of host bacterial interactions. However, available data reveal that the competitive advantage conferred by initiating RIPK1- and/or RIPK3-dependent cell death may lie with the host or the bacteria, depending on the identity of the bacteria. For example, cell death may function as part of the productive innate immune responses to mitigate infection; alternatively, cell death may be induced by bacteria to eliminate immune cells and trigger massive tissue damage and thus may be targeted for therapeutic intervention. A broader understanding of the bacterial contexts regulating RIPK1- and/or RIPK3-dependent cell death and their underlying mechanisms remains to be clarified.

3 Epithelial Barrier Function

The intestinal epithelia and skin are first-line defenses against infection by commensal and pathogenic microbial flora (Kumar et al. 2010; Bonnet et al. 2011; Kaser et al. 2010; Macdonald and Monteleone 2005). Destruction of critical epithelial barriers predispose to microbial infection and the generation of tissue inflammation. The literature has illuminated crucial roles for RIPK1 and RIPK3 in maintaining barrier tissues. For example, kinase-independent function of RIPK1 has been found to be essential in preserving intestinal epithelial integrity and skin homeostasis. Conversely, kinase activity of RIPK1 and activation of RIPK3-dependent necroptosis have been associated with epithelial tissue destruction. The ensuing discussion explores these opposing roles of RIPK1 and RIPK3 in barrier tissue biology.

3.1 *Intestinal Epithelia*

3.1.1 Kinase-Independent Functions of RIPK1

Two groups have reported roles for RIPK1 in maintaining epithelial integrity using independently generated mouse models of RIPK1 deletion in intestinal epithelial cells ($RIPK1^{IEC}$). Grossly, weight loss and diminished survival was observed in mice lacking epithelial-specific RIPK1. Histological evaluation of gastrointestinal tissue revealed atypical cell death of intestinal epithelial cells and features of inflammatory injury, including leukocyte infiltration and abnormal tissue architecture (Dannappel et al. 2014; Takahashi et al. 2014).

Loss of RIPK1 in intestinal epithelia sensitized mice to microbial injury. Takahashi and colleagues reported that broad-spectrum antibiotic treatment prevented weight loss and improved survival in animals. Antibiotics treatment also protected against cellular apoptosis in intestinal tissue as well as signs of local tissue and systemic inflammation, shortened colon length and splenic enlargement, respectively. Conversely, Dannappel and colleagues found that antibiotic treatment was ineffective at reversing histological findings of cellular apoptosis; however, these differences may be attributed to variations in antibiotic regimen and treatment schedule.

RIPK1 deletion in intestinal epithelia was associated with increased sensitivity to inflammation-associated cell death. TNF-α treatment resulted in increased apoptosis of RIPK1^{IEC} organoids (Takahashi et al. 2014). In addition, deletion of RIPK1 was associated with increased death induced by innate immune ligand, polyinosinic: polycytidylic acid [Poly(I:C)], as well as the cytokines, interferon-β (IFN-β), and interferon-γ (IFN-γ) (Kaiser et al. 2014; Dillon et al. 2014). Congruently, deletion of TNF Receptor (TNFR$^{-/-}$) prolonged survival and ameliorated intestinal apoptosis in RIPK1^{IEC} mice. Although Takahashi and colleagues appreciated improved survival in RIPK1^{IEC} mice upon deletion of MYD88, a key adaptor in innate immune inflammatory signaling and downstream of multiple Toll-like receptor family members, these findings could not be affirmed by Dannappel et al. Nevertheless, together, these observations suggest that deletion of RIPK1 in intestinal epithelia increased tissue sensitivity to death by innate immune signals and cytokines.

Similarly, whole-body RIPK1 deletion results in unbridled postnatal inflammation and lethality. Notably, embryonic development is unaffected by RIPK1 deletion, highlighting the importance of RIPK1 in post-embryonic life, when the body is exposed to a variety of external factors, in particular, skin and gut colonization by microbial flora (Dillon et al. 2014; Kaiser et al. 2014; Rickard et al. 2014; Kelliher et al. 1998).

The role for RIPK1 is not exclusively limited to early development as acute deletion of RIPK1 in intestinal epithelia of adult mice also resulted in rapid death associated with apoptosis of intestinal epithelia (Takahashi et al. 2014). Moreover, the protective role of RIPK1 in intestinal epithelia was credited to its kinase-independent function or scaffold function because mouse models of RIPK1 kinase-inactivation have normal survival and fail to exhibit pathologic features associated with RIPK1 deletion (Takahashi et al. 2014; Dannappel et al. 2014).

RIPK1 was previously thought to regulate pro-inflammatory and pro-survival signaling by the regulation of downstream transcription factor NFκB (Cusson-Hermance et al. 2005; Meylan et al. 2004). However, counter to the dogma, Takahashi and Dannappel both reported that NFκB signaling was not impaired in RIPK1^{IEC} epithelia. Specifically, inhibitor of kappa-B (IκB) degradation was unchanged in RIPK1^{IEC} organoids stimulated with TNF-α.

Epithelial cell death and inflammatory sequelae were attributed to caspase-8- and FADD-dependent apoptosis and/or RIPK3-dependent necroptosis. These insights were gleaned using intestinal epithelia-specific knockouts of caspase-8 (Casp8^{IEC}) or FADD (FADD^{IEC}) and RIPK3 deletion mutant mice in conjunction with RIPK1^{IEC} mice. Caspase-8^{IEC}/RIPK1^{IEC} mice were completely protected against early lethality and histological features of intestinal pathology (Takahashi et al. 2014). Although FADD depletion in intestinal epithelia (FADD^{IEC}) of RIPK1^{IEC} mice protected against histological features of intestinal pathology in a dose-dependent manner, these mice still exhibited diminished survival. Conversely, RIPK1^{IEC}/RIPK3$^{-/-}$ were not protected from inflammatory changes in the intestinal tissue and had diminished survival. Nevertheless, inflammatory histopathology and survival were completely rescued in RIPK1^{IEC}/FADD^{IEC}/RIPK3$^{-/-}$indicating that both FADD and RIPK3 were important for pathologic changes in the tissue in the absence of RIPK1 (Dannappel et al. 2014).

Dannappel and colleagues also suggested that RIPK1 maintained intestinal epithelial integrity by preserving the pro-survival or anti-apoptotic proteins, cIAP1, TRAF-2, and c-FLIP. TNF-α stimulation resulted in rapid degradation of cIAP1, TRAF-2, and c-FLIP in RIPK1 knockout (RIPK1$^{-/-}$) but not RIPK1 kinase-inactive MEFs. Moreover, tamoxifen-induced deletion of RIPK1 in organoid cultures was associated with reduced expression of these anti-apoptotic proteins and rapid organoid death (Dannappel et al. 2014).

3.1.2 Kinase-Dependent Regulation by RIPK1 and the Role of RIPK3

Loss of caspase-8 or death-receptor adaptor protein, Fas-associated death domain (FADD), has been shown to sensitize cells to necroptotic cell death (Zhang et al. 2009; He et al. 2009; Osborn et al. 2010; Vandenabeele et al. 2010b; Kim and Li 2013). Several groups have analyzed the consequences of genetic deletion of these factors on the integrity of the intestinal epithelium.

In 2011, Welz et al. reported that FADD prevents RIPK3-dependent intestinal epithelial cell loss in mice. Mice generated with a deletion of FADD in intestinal epithelial cells (FADD^{IEC}) displayed diminished survival, reduced weight, and diarrhea, suggesting intestinal disease (Welz et al. 2011). Evaluation of gastrointestinal tissue revealed necrotic loss of intestinal epithelia or enterocytes, enteric and colonic inflammation, and destruction of antimicrobial Paneth cells of crypts found in the small intestine. Conspicuously, FADD^{IEC}/RIPK3$^{-/-}$ mice developed normally and were protected from intestinal epithelial cell loss, Paneth cell loss, and signs of inflammation, indicating a role for RIPK3-dependent intestinal epithelial cell loss in the generation of gastrointestinal inflammatory disease (Welz et al. 2011). Moreover, intestinal epithelial inactivation of CYLD, a cellular deubiquitinase reported to promote RIPK1- and RIPK3-dependent necroptosis, protected animals from developing gastrointestinal inflammation (Moquin et al. 2013). Significantly, antibiotic treatment, germ-free rearing, or concomitant deletion of MYD88 or TNF ameliorated or abolished signs of intestinal inflammatory disease, emphasizing the importance of gut flora and innate immune responses in the development of gastrointestinal disease upon loss of intestinal epithelia (Welz et al. 2011). Collectively, these observations suggest that activation of RIPK3-dependent necroptosis may promote intestinal epithelial cell loss and bacterial-associated gastrointestinal inflammation.

RIPK3-dependent intestinal tissue damage was also observed upon the genetic deletion of caspase-8. Weinlich and colleagues administered oral tamoxifen gavage to Rosa-CRE-expressing mice that were bred to casp8 $^{fl/fl}$ mice to generate mice lacking caspase-8 in adult gastrointestinal tissue (Casp8^{GI}) (Weinlich et al. 2013). In an alternative approach, Günther and colleagues generated mice lacking caspase-8 in intestinal epithelial cells (Casp8^{IEC}) by breeding caspase-8 floxed (Casp8 $^{fl/fl}$) mice with mice expressing CRE-recombinase under the regulation of intestinal epithelial-specific *villin* promoter.

Acute deletion of caspase-8 in adult mice promoted rapid weight loss and diminished survival (Weinlich et al. 2013). Casp8^{GI} mice also manifested with histological features of intestinal inflammation, marked by cell death of enterocytes, tissue inflammation, and infiltration of immune cells. Importantly, concomitant deletion of RIPK3 protected mice from intestinal disease noted in Casp8^{GI} mice. Casp8^{GI}/RIPK3$^{-/-}$ mice maintained body weight and had normal survival, fewer dying enterocytes, and intact gastrointestinal tissue architecture, indicating that intestinal damage was likely a consequence of RIPK3-dependent enterocyte death in Casp8^{GI} mice (Weinlich et al. 2013). Similarly, Casp8^{IEC} mice were found to have spontaneous inflammatory lesions in the terminal ileum, marked by bowel wall thickening, loss of intestinal crypts, and increased cellularity, suggesting immune cell infiltration. Notably, Casp8^{IEC} mice had increased epithelial cell death associated with necrotic features. Authors also observed loss of specialized epithelial cells, specifically goblet and Paneth cells, which protect enteric epithelia and aid immunity by producing mucus and secreting antimicrobial peptides, respectively (Günther et al. 2011). Not surprisingly, intestinal epithelia of these mice had increased sensitivity to TNF-α-induced necroptosis. Intravenous TNF-α resulted in high lethality associated with severe destruction of the small bowel and an increased number of dying epithelial cells that were negative for cleaved caspase-3. Significantly, inhibition of RIPK1 kinase-dependent cell death using Nec-1 blocked TNF-α-induced lethality and destruction of the small bowel (Günther et al. 2011).

Together, these studies demonstrate that activation of necroptosis in intestinal epithelia leads to loss of intestinal barrier integrity and induction of tissue inflammation. These phenotypes may be associated with commensal bacterial-induced innate immune responses. Notably, the data also suggest that inappropriate activation of RIPK3-dependent signaling in human intestinal epithelia may be linked to chronic inflammatory diseases of the bowel, namely Crohn's disease. Immunohistochemical analysis of tissue samples from human patients revealed increased RIPK3 expression in Paneth cells of the distal small bowel. Crohn's disease tissue specimens also had decreased numbers of Paneth cells and increased Paneth cell death with necrotic features as determined by electron microscopy. Moreover, Paneth cells from tissue biopsies of patients with Crohn's disease were susceptible to TNF-α-induced cell death that could be blocked with Nec-1 (Günther et al. 2011). These data suggest that aberrant activation of necroptosis and subsequent loss of barrier function may be important factors underlying chronic inflammation in the gut.

3.2 Skin

3.2.1 Kinase-Independent Functions of RIPK1

Similar to the intestinal epithelia, RIPK1 in keratinocytes and epidermal tissue is important for homeostasis of the skin. RIPK1-deficient primary keratinocytes have increased susceptibility to TNF-α-induced apoptosis; however, in contrast to intestinal epithelia, injury in vivo entirely dependent on necroptosis as RIPK3 deletion protects against pathological features of skin inflammation. Moreover, ablation of RIPK1 in epithelia ($RIPK1^E$) results in a pro-inflammatory phenotype with increased epidermal thickness and inflammation (Dannappel et al. 2014). Localized inflammation of the skin may disrupt integrity of the organ, thereby increasing susceptibility to infection and exacerbating inflammatory pathology. This theory is supported by the observation that $RIPK^E$ mice also lacking the Toll-like receptor adaptor protein TRIF were partially protected from inflammatory injury of the skin, likely in part due to decreased activation of innate immune pathways by microbial flora (Dannappel et al. 2014). Additionally, much of the inflammatory injury observed upon epithelial deletion of RIPK1 may be attributed to increased sensitivity to TNF-α signaling as was observed in intestinal epithelia. $RIPK1^E/TNFR^{-/-}$ prevents inflammatory features in the skin, including epidermal thickness (Dannappel et al. 2014). Analogous to observations made in intestinal epithelia, anti-apoptotic ubiquitin ligase, cIAP1, is rapidly diminished following TNF-α exposure to $RIPK1^E$ to primary keratinocytes (Dannappel et al. 2014). Lastly, inflammatory phenotypes associated with RIPK1 deletion in the epithelia appear to be independent of kinase activity of RIPK1 as kinase-inactive models of RIPK1 have not been found to manifest with spontaneous inflammation of the skin (Polykratis et al. 2014; Berger et al. 2014).

3.2.2 The Role of RIPK3

Deletion of FADD in epidermal keratinocytes resulted in development of RIPK3-dependent skin lesions and early death (Bonnet et al. 2011). Histologically, these mice displayed patchy pathologic skin signs, marked by epidermal hyperplasia and thickening, keratinocyte death, and immune cell accumulation. Importantly, keratinocyte death appeared to be caspase independent, as a large fraction of dying keratinocytes did not contain active caspase-3 but showed necrotic morphology by electron microscopy. Similar to FADD deficiency in intestinal epithelium, $FADD^E/RIPK3^{-/-}$ mice neither exhibited early lethality nor developed skin lesions during development and into adulthood. In fact, $FADD^E/RIPK3^{-/-}$ mice had normal skin, with normal epidermal thickness and an absence of dying keratinocytes and pathologic inflammatory infiltrates, indicating that inflammatory disease observed in $FADD^E$ mice was entirely dependent on RIPK3-mediated cell death. Concordantly, $FADD^E$ mice with concomitant deletion of TNF-α-receptor or MYD88 or an inactivating mutation of CYLD had delayed onset of inflammatory skin lesions, stressing that compromise of barrier function and microbial-associated innate immune activation may be responsible for the disease pathogenesis (Bonnet et al. 2011).

Constitutive loss of caspase-8 in epidermal keratinocytes produced an inflammatory skin disorder with epidermal hyperplasia, dermal inflammatory cell infiltration, and premature lethality in mice (Kovalenko et al. 2009). These findings were corroborated by studies in which acute deletion of caspase-8 in adult murine skin by application 4-hydroxytamoxifen to Rosa-CRE, Casp8 $^{fl/fl}$ mice resulted in local tissue inflammation and damage (Welz et al. 2011). Importantly, 4-hydroxytamoxifen treatment of Rosa-CRE, Casp8 $^{fl/fl}$, RIPK3$^{-/-}$ mice resulted in minimal dermal inflammation, epidermal hyperplasia, and keratinocyte death (Welz et al. 2011). In accord, RIPK3-dependent signaling promoted skin inflammation and injury upon deletion of caspase-8 in epidermal tissue. Although the role of RIPK1 has not been specifically established in these studies, it could be anticipated that RIPK1 kinase activity may be important as an inducer of RIPK3-dependent cell death.

3.3 *Summary*

Loss of essential barrier cells lends the organism to infection by commensal microbiological flora, tissue injury, and diminished survival. The studies described above highlight an important new role for kinase-independent function of RIPK1 in maintaining epithelial homeostasis of the intestinal tract as well as in the skin. RIPK1 serves this function by two apparent means that may be fundamentally intertwined. First off, the presence of RIPK1 directly prevents uncontrolled activation of caspase-8 and FADD-associated apoptosis and RIPK3-dependent necroptosis at the epithelial surface. Secondly, RIPK1 promotes a survival-like state in the presence of pro-inflammatory and pro-injury signals such as TNF-α, IFNb, and other ligands associated with infection. Indeed, dual roles of RIPK1 in both activating and inhibiting necroptotic cell death have been described in the literature (Kearney et al. 2014; Orozco et al. 2014). Furthermore, evidence reported here also demonstrates that aberrant or inappropriate activation of necroptosis, driven by RIPK1-kinase function or RIPK3, can result in loss of barrier function in the skin and intestinal epithelia. Deletion of caspase-8 or FADD in barrier tissue sensitizes cells to necrotic death and chronic inflammatory disease that may be attributed to microbial activation of the innate immune host response.

4 Inflammation

Inflammation is an integral part of the host response (Kumar et al. 2010). RIPK1 and RIPK3 are well recognized for inducing a pro-inflammatory form of cell death; however, accumulating evidence suggests that in some cases, these enzymes may also regulate inflammation independent of necroptosis. It should be noted that this distinction may be difficult to draw in many cases due to a highly intertwined nature of cell death and inflammation. The ensuing discussion highlights emerging, but still poorly characterized roles for RIPK1 and RIPK3 in amplifying the host inflammatory response following cytokine stimulation, activation of innate immune receptors, and microbial infection.

4.1 RIPK1

TNF-α is well recognized for regulating RIPK1 kinase-dependent cell death; however, TNF-α is also a major pro-inflammatory cytokine that promotes nuclear factor kappa B (NFκB)-associated inflammation (Christofferson et al. 2012; Vandenabeele et al. 2010a; Hitomi et al. 2008; Kelliher et al. 1998). Kinase-independent function of RIPK1 has been linked to NFκB activation in response to TNF-α. Expression of kinase-inactive RIPK1 restores defective TNF-α-induced NFκB activation in $RIPK1^{-/-}$ Jurkat cells (Ting and Pimentel-muiffios 1996). Similarly, analysis of nuclear extracts from embryonic fetal liver-derived transformed pre-B cells demonstrated decreased NFκB–DNA binding in electrophoretic mobility shift assays in $RIPK1^{-/-}$ cells (Kelliher et al. 1998). Conspicuously, reports have found that TNF-α-induced NFκB activation is associated with covalent modification of RIPK1 (Zhang et al. 2000). These findings highlight an important role for kinase-independent function of RIPK1 in promoting TNF-α-induced NFκB-driven inflammation.

Roles for RIPK1- dependent NFκB regulation have also been uncovered downstream of innate immune receptors. In 2004, Tschopp and colleagues noted that RIPK1 promotes NFκB activation downstream of TLR3, the pathogen recognition receptor (PRR) for sensing viral double-stranded RNA (Meylan et al. 2004). Activation of NFκB pathway, assessed by phosphorylation of inhibitor of kappa B (IκB), is impaired in MEFs lacking RIPK1 following stimulation with TLR3 agonist, Poly(I:C). Exogenous protein expression in HEK293T cells demonstrated that RIPK1 associates with TLR3 adapter protein, TRIF, by common RHIM-domain interactions. Furthermore, expression of TLR3, TRIF, and RIPK1 in HEK293T cells resulted in their co-immunoprecipitation, suggesting that the three factors may complex following receptor activation (Meylan et al. 2004). Similar studies in MEFs demonstrated that RIPK1 is required for NFκB activation following ligation of TLR4 with Gram-negative bacteria associated pathogen-associated molecular pattern, lipopolysaccharide (LPS) (Cusson-Hermance et al. 2005). Importantly, LPS-induced NFκB activation also requires TRIF (Cusson-Hermance et al. 2005). Finally, studies found that TRIF-dependent NFκB activation manifested in correlation with increased phosphorylation and ubiquitylation of RIPK1, suggesting importance of post-translational modification in RIPK1-dependent NFκB activation. Together, these results find that RHIM-domain interactions likely facilitate a role of RIPK1 in NFκB activation downstream of innate immune receptors, TLR3 and TLR4.

Apart from NFκB signaling, kinase-independent scaffold properties of RIPK1 have also been recognized in PRR activation of the interferon-inducing transcription factor and interferon regulatory factor 3 (IRF3) (Rajput et al. 2011). RIPK1 localizes to mitochondria in association with cytosolic RNA/DNA sensor RIG-I following viral infection with Sendai virus (SeV). These events occur in coordination with IRF3 activation and expression of downstream genes. Knockdown of RIPK1 impairs SeV-induced IRF3 activation. Moreover, MEFs reconstituted with a non-ubiquitylatable form of RIPK1 at K377 have deficits in SeV-induced IRF3 activation, indicating that ubiquitin conjugation of RIPK1 at K377 is necessary for activation of IRF3 (Rajput et al. 2011). Remarkably, in spite of an abundance of evidence linking ubiquitylation of RIPK1 to NFκB activation, these studies found no changes in NFκB activation following SeV infection. Accordingly, it appears as though RIPK1 may be a versatile regulator in innate immune pathways, and precise action of RIPK1 is determined in a context-dependent manner.

Few studies have ventured to uncover roles for RIPK1-dependent inflammation in bacterial infection models. However, recent investigation of RIPK1 in *Yersinia* infection discovered that RIPK1$^{-/-}$ fetal liver macrophages carried defects in LPS as well as *Yersinia*-induced secretion of IL-6, a pyrogen that directs systemic inflammatory responses. Additionally, caspase-1 cleavage was diminished in RIPK1$^{-/-}$ fetal liver macrophages infected with *Yesinia* species, indicating that RIPK1 may be required for caspase-1-driven inflammasome activation and consequently IL1β and IL-18 secretion (Weng et al. 2014). It is unclear as to whether IL6 secretion and/or caspase-1 cleavage requires RIPK1 kinase activity as authors did not use inhibitors to evaluate kinaseactivity in these studies. Nevertheless, these observations identify roles for RIPK1 in multiple pro-inflammatory pathways in a model of Gram-negative bacterial infection.

4.2 Kinase-Dependent Inflammatory Functions of RIPK1

While kinase activity of RIPK1 is commonly associated with cell death, a number of recent reports suggest that inflammasome activation by RNA viruses, including VSV, SeV, and influenza (Flu), may occur in a RIPK1 kinase-dependent manner. The inflammasome is a molecular complex comprised of the receptor NLRP3 and caspase-1 that is responsible for the maturation and secretion of pro-inflammatory cytokines, IL1β and IL-18. Nec-1 or knockdown of RIPK1 reduced IL1β secretion in bone marrow-derived macrophages (BMDMs) infected with VSV (Wang et al. 2014a).This regulation was also found to require RIPK3 as RIPK3$^{-/-}$ BMDMs also had similar defects in inflammasome activation following RNA virus infection. In a physiologic system, RIPK3$^{-/-}$ mice produced lower levels of IL1β and IL-18 in response to infection with RNA viruses as well (Wang et al. 2014a). Surprisingly, MLKL was not required for inflammasome activity, as determined using MLKL knockout (MLKL$^{-/-}$) BMDMs, suggesting that RIPK1 and RIPK3 regulate inflammasome activation independent of necroptosis pathway. Moreover, VSV-induced minimal death in target cells indicating that RIPK1 and/or RIPK3 were likely not regulating alternative death pathways such as pyroptosis. Mechanistic analysis revealed that RIPK1 and RIPK3 complexed with mitochondrial fission protein, DRP1, and were required for translocation of DRP1 to mitochondria following infection. DRP1 was required for RNA virus-induced mitochondrial fission, mitochondrial aggregates, ROS generation, and inflammasome activation as well. Poly(I:C), double-stranded RNA mimic, was found to induce NLRP3-dependent inflammasome activation in a RIPK1, RIPK3, and DRP1-dependent manner, indicating that viral nucleic acids may be the ignition for RIPK1 kinase and RIPK3- dependent inflammasome activation (Wang et al. 2014a).

Intrinsic kinase activity of RIPK1 may also underlie certain tissue-specific inflammatory processes. $Ptpn6^{spin}$ mice harboring a Tyr208Asn mutation in the non-receptor protein tyrosine phosphatase Src-homology region 2 domain-containing phosphatase-1 (SHP-1) develop footpad inflammation that is RIPK1 kinase dependent (Lukens et al. 2013). Studies revealed that aberrant production of the cytokine IL1α exacerbates inflammation in $Ptpn6^{spin}$-mediated disease. Nec-1 diminished IL1α expression and ameliorated footpad inflammation in $Ptpn6^{spin}$ mice. Additionally, reconstitution of WT mice with $Ptpn6^{spin}/RIPK1^{-/-}$ fetal liver (hematopoietic precursor) cells abrogated the inflammation produced by transfer of $Ptpn6^{spin}/RIPK1^{+/+}$ fetal liver cells. Notably, $Ptpn6^{spin}/NLRP3^{-/-}$, $Ptpn6^{spin}$/Caspase-$1^{-/-}$, or $Ptpn6^{spin}/RIPK3^{-/-}$ mice were not protected from $Ptpn6^{spin}$-inflammatory disease. These findings suggest a critical role for kinase function of RIPK1 in hematopoietic cells to promote IL1α-dependent inflammation that is independent of inflammasome components and RIPK3 (Lukens et al. 2013).

Similarly, mouse models of RIPK1 and RIPK3 kinase activation, such as tissue-specific caspase-8 deletion in dendritic cells, keratinocytes, and or intestinal epithelium, have spawned inflammatory disease in mice (Kovalenko et al. 2009; Günther et al. 2011; Cuda et al. 2014). Although it may be unclear as to whether inflammation in some of these models is occurring independent of necroptosis, Cuda and colleagues observed that deletion of caspase-8 in dendritic cells (DCs) generated an autoimmune condition in mice that could not be attributed to impaired cell survival or a RIPK3-mediated processes. Caspase-8 deletion in DCs crossed to $RIPK3^{-/-}$ mice was not protected from inflammatory pathology. Rather, authors observed that systemic inflammation was partially dependent on MYD88 and also found that Nec-1 abrogated Toll-like receptor activation-induced expression of TNF-α, IL-6, and IL1β (Cuda et al. 2014). These accounts suggest roles for kinase function of RIPK1 in regulating Toll-like receptor-induced cytokine expression in dendritic cells.

The recent availability of RIPK1 kinase-inactive mouse models has unveiled significance of RIPK1 catalytic activity in directing inflammation. For example, RIPK1 kinase-inactive mice are protected from TNF-α-induced hypothermia and shock (Polykratis et al. 2014). Another group evaluated kinase function of RIPK1 in the absence of SHARPIN, a component of the linear ubiquitin assembly complex (LUBAC). SHARPIN deletion results in gross inflammatory pathology, dermatitis, and diminished organism survival. Authors observed that crossing their independently generated RIPK1 kinase-inactive mice to SHARPIN-deficient mice protected animals from inflammatory pathology and prolonged survival (Berger et al. 2014).

4.3 RIPK3

Work spearheaded by Francis Chan and colleagues has marked RIPK3 as an important component of the host inflammatory response to injury and infection. In one study, the group reported impaired inflammation-associated repair in RIPK3$^{-/-}$ mice in the dextran sodium sulfate (DSS) model of colitis (Moriwaki et al. 2014). Specifically, RIPK3 in hematopoietic cells is required for protection against DSS colitis; RIPK3$^{-/-}$ mice with WT bone marrow lost less body weight and had reduced inflammation (Moriwaki et al. 2014). RIPK3$^{-/-}$ mice had reduced circulating levels of IL1β and IL23, and resupplementation of these cytokines mitigated the extent of colitis. In concordance, bone marrow-derived dendritic cells (BMDCs) had defects in LPS-induced expression of cytokines, including TNF-α, IL1b, IL23, and monocyte chemotactic protein-1 (MCP-1) (Moriwaki et al. 2014). It is unlikely that these observations are linked to the differences in cell death or the release of damage-associated molecular patterns as RIPK3$^{+/+}$ and RIPK3$^{-/-}$ BMDCs did not undergo death in response to LPS. Rather, these effects were attributed to defects in the generation of ROS and nuclear localization of subunits RelB and p50 in the NFκB pathway (Moriwaki et al. 2014). Similar defects in inflammatory cytokine expression were not observed in BMDMs suggesting that RIPK3 may have a broad functional range that can be employed in a context or cell-specific manner.

Analyses in BMDCs demonstrated that RIPK3 is required for caspase-8 and caspase-1 dependent IL1β processing and secretion. Authors observed that LPS-induced RIPK3 complex formation with caspase-8 in BMDCs. Conspicuously, both caspase-8 cleavage as well as IL1β secretion were decreased in LPS-stimulated RIPK3$^{-/-}$ BMDCs suggesting that RIPK3 directs LPS-induced IL1β secretion by facilitating maturation or activation of caspase-8 (Moriwaki et al. 2015). Kinase activity of RIPK3 appeared to be dispensable as RIPK3 kinase-inactive BMDCs did not display a deficit in IL1β production (Moriwaki et al. 2015). Authors also observed that kinase-independent function of RIPK1 may be required as lentiviral CRE-mediated deletion of kinase-inactive RIPK1 in BMDCs reduced LPS-induced caspase-8 maturation; however, no defects were observed in absence of CRE. Paradoxically, RIPK3 kinase inhibitor, GSK'872, augmented LPS-induced IL1β production, suggesting that this regulation may not be controlled by the catalytic activity of RIPK3 per se, but may be influenced by the conformation of the kinase domain. Consistent with this model, GSK'872 enhanced the recruitment of RIPK1 to RIPK3 (Moriwaki et al. 2015).

4.4 Summary

Inflammation is a crucial part of the host response during infection. Notably, RIPK1 and RIPK3 have been identified in critical roles regulating pro-inflammatory transcription factors, cytokine synthesis, and cytokine secretion in response to immunogenic ligands. Moreover, RIPK1 kinase activity and RIPK3 have been linked to tissue-specific and systemic inflammation. As discussed, in some instances, inflammatory signaling by these proteins can be clearly separated from cell death. Accordingly, these factors may emerge as potential therapeutic targets not only in pathologic settings that are driven by significant cell death and tissue loss, but also in disease states that are primarily inflammatory in nature.

5 RIPK1 and RIPK3 as Therapeutic Targets in Light of New Innate Immune Roles

RIPK1 and RIPK3 emerged as potential targets in many human pathologies involving necrotic cell death in models of stroke, tissue infarction, atherosclerosis, ischemia–reperfusion injury, pancreatitis, inflammatory bowel disease, and others (Linkermann and Green 2014; You et al. 2008; Degterev et al. 2005). The emerging roles of these proteins in innate immunity need to be considered when evaluating therapeutic potential of targeting RIPK1 and RIPK3. However, involvement of RIPK1 and RIPK3 in responses to pathogens per se should not necessarily diminish enthusiasm for developing RIPK-targeting therapies. In particular, while RIPK1 and RIPK3 may regulate robustness of innate immune responses, it is unclear whether inhibition of these pathways may truly change the outcome of infection as opposed to just changing the kinetics of cell death and, ultimately, the death of the organism. It should also be noted that, thus far, neither RIPK1 nor RIPK3 have emerged as critical players in any of the paradigms involving reactivation of the dormant pathogens. On the other hand, in cases of acute infection or sepsis, in which cell death and inflammation are the reason for high mortality, these pathways may be targeted to transiently attenuate the intensity of the host responses to improve patient outcomes, especially when longer term therapeutic options are available to ultimately eradicate pathogens. Needless to say, better understanding of the balance of RIPK1 vs RIPK3 contributions using recently developed genetic and pharmacologic tools will also be important. Current evidence suggests overlapping functions of these regulators in a number of paradigms, including pathogen-induced necroptotic cell death, as well as possibly distinct roles in response to particular pathogens. For example, RIPK3 may be the primary mediatory of certain virus-induced responses, whereas RIPK1 is responsible for YopJ-induced apoptosis by *Yersinia* (Table 1) (Upton et al. 2012; Kaiser et al. 2013; Moriwaki et al. 2014; Lamkanfi and Dixit 2010; Weng et al. 2014; Philip et al. 2014).

6 Discussion

Emerging evidence presented here illuminates roles for RIPK1 and RIPK3 in fortifying host immunity against certain pathogens. Specifically, RIPK1 and RIPK3 appear to play a role in the host response to mitigate viral dissemination and promote host survival in viral infection. RIPK1 and RIPK3 may also play multiple, complex roles in bacterial pathogenesis as well as contribute to other facets of innate immunity, including epithelial barrier homeostasis, and inflammation occurring independent of cell death.

RIPK1- and/or RIPK3-dependent cell death is an important escape tool for cells infected by viral species, which depend on the survival of host cells to propagate. As noted, viral pathogens are armed with caspase or apoptosis inhibitors to subvert the cell death defense and promote microbial dissemination. Only a few pathogens, such as HSV and CMV, have been analyzed with respect to regulation of necrotic cell death and, intriguingly, were found to be equipped with mechanisms to disrupt pro-necrotic signaling. Thus, the question remains as to whether inhibition of necroptosis represents a common element of viral infection, akin to targeting apoptosis, or inhibition of necroptosis reflects a recent evolutionary addition to viral–host interactions and, thus, may only be observed among a limited number of viral pathogens (Omoto et al. 2015; Guo et al. 2015; Huang et al. 2015; Wang et al. 2014b).

It is necessary to clarify that the utility of caspase or apoptosis inhibitors in viral-induced RIPK1- and/or RIPK3-dependent cell death is largely assumed. In fact, vaccinia virus is the only available viral infection model in which caspase inhibition has been shown to be required to induce necrotic cell death (Li and Beg 2000). Likewise, the occurrence of caspase inhibition in RIPK1 and RIPK3 activation associated with bacterial infection has not been examined (Faherty and Maurelli 2008). In the examples summarized here, bacterial-induced RIPK1 and/or RIPK3 activation is linked to the presence of specific bacterial effector proteins and/or toxins, namely *Yersinia* and YopJ, *E. coli* and Nleb1, and *Clostridium* and CPB. Programmed necrosis in *SA* infection is associated with multiple toxins, including Hla and leukotoxins. RIPK1 and/or RIPK3 activation in *Salmonella* and *M. tb* infections is associated with host cytokine responses, IFN-I and TNF-α, respectively. In a few of these systems, RIPK1-dependent apoptosis, rather than necrosis has been reported. Thus, another key question remains as to whether activation of RIPK1 and/or RIPK3 signaling is generally a consequence of pathogen-dependent caspase inhibition or represents a cellular response independent of caspase regulation. It should also be noted that caspase inhibition or inhibition of other RIPK1 and/or RIPK3 modulators, such as cIAP ubiquitin ligases or Tak1 kinase, appears to be required for activation of RIPK1 and/or RIPK3 by many of the known triggers (i.e., TNF family, IFNs, TLR agonists) in vitro (Christofferson et al. 2014). However, it is similarly not required in many of the in vivo models studied to date, some of which clearly involve the same triggers. Thus, fully understanding the mechanism(s) silencing RIPK1 and RIPK3 signaling in vitro, but likely failing to do so efficiently in vivo, appears to be another critical step in fully revealing the roles of RIPK1 and RIPK3 proteins in innate immune regulation.

Although a significant body of evidence indicates that RIPK1- and RIPK3-dependent cell death pathways are beneficial for limiting the propagation of viral pathogens, the evidence for RIPK1 and RIPK3 in bacterial pathogenesis is equivocal. It has been proposed that bacterial-induced necrosis pathways may serve to restrain the host immune response and promote leukopenic states (Lamkanfi and Dixit 2010). This viewpoint suggests that bacterial species, unlike viruses, may not be under selective pressure to inhibit cell death because these pathogens are able to grow and replicate independent of the host cell. This perspective is consistent with observations of *SA*-induced necrotizing pneumonia or *Salmonella* and PFP-induced death discussed here. Conversely, evidence from studies exploring the pathogenesis of *Yersinia* species suggests that RIPK1-dependent cell death pathways aid in host survival. These pathogens differ in a variety of respects; however, notably, the literature suggests that *Yersinia* may have an intracellular phase that is a feature of its pathogenesis (Pujol et al. 2009; Pujol and Bliska 2003, 2005). Under this circumstance, host cells provide nourishment for invading bugs and activation of programmed cell death limits access to this resource. Accordingly, RIPK1 and/or RIPK3-dependent cell death in bacterial infection may not be a generalizable process; however, appreciating infection-specific impact of RIPK1- and/or RIPK3-dependent cell death may be of value in managing infectious disease clinically. For example, although physiologic RIPK1 and/or RIPK3-dependent cell death may be effective in limiting certain bacterial infections, in other instances, therapeutic targeting of RIPK1- and/or RIPK3-dependent cell death may aid in limiting pathogen-induced tissue damage and improving patient outcomes.

Apart from roles in pathogen-induced cell death, kinase-independent and kinase-dependent functions of RIPK1 and RIPK3 have also ascended as important elements in epithelial barrier integrity. Kinase-independent function of RIPK1 sequesters pro-apoptotic activity of caspase 8–FADD complex and pro-necrotic function of RIPK3. Loss of surface epithelia is associated with infection-associated tissue damage, inflammation, and diminished organism survival, signifying that the role of RIPK1 in maintaining epithelial integrity is an important first-line defense against microbial species. Conversely, activation of RIPK1 and RIPK3 kinase-dependent signaling, through the loss of FADD or caspase-8, triggers loss of epithelial barrier function, and the role of this process in a variety of inflammatory pathologies remains to be elucidated from the therapeutic prospective.

Lastly, RIPK1 and RIPK3 are gaining repute as regulators of pro-inflammatory processes independent of cell death. This summary of available literature highlights roles for RIPK1 and RIPK3 in regulating inflammatory pathways in response to cytokine signals, innate immune ligands, and microbial species. However, direct evidence to implicate the role of RIPK1- and/or RIPK3-dependent inflammation in microbial pathogenesis and organism survival in vivo is currently unavailable. Importantly, its emergence would surely confirm this new role in innate immunity for these proteins. Understandably, scientists may shy away from these ventures because of the specter of underlying cell death in experimental models and the clearly established pro-inflammatory nature of cell death processes; however, these confounding factors can be addressed experimentally. For example, evaluating RIPK1- and RIPK3-dependent inflammation on a $MLKL^{-/-}$ background may provide interesting insights into necroptosis-independent functions of RIPK1 and RIPK3. Alternatively, infection-associated RIPK1- and RIPK3-dependent function may be evaluated in models of cytokine depletion and inflammatory attenuation. Indeed, efforts are underway to discriminate RIPK1/3-dependent inflammation from cell death (Kang et al. 2014). Importantly, it is necessary to recognize that these answers will be important in ultimately addressing the significance of RIPK1- and/or RIPK3-dependent inflammation in the host response.

References

Albrecht ME (2014) Epidemiology, clinical manifestations, and diagnosis of genital herpes simplex virus infection. Uptodate http://www.uptodate.com/contents/epidemiology-clinical-manifestations-and-diagnosis-of-genital-herpes-simplex-virus-infection?source=search_result&search=HSV-2&selectedTitle=1 ~ 68

Alpuche-Aranda CM, Racoosin EL, Swanson JA, Miller SI (1994) Stimulate macrophage macropinocytosis and persist within spacious phagosomes. J Exp Med 179(2):601–608

Autheman D, Wyder M, Popoff M, D'Herde K, Christen S, Posthaus H (2013) Clostridium perfringens beta-toxin induces necrostatin-inhibitable, calpain-dependent necrosis in primary porcine endothelial cells. PloS One 8(5):e64644. doi:10.1371/journal.pone.0064644. http://www.pubmedcentral.nih.gov/articlerender.fcgi?artid=3667183&tool=pmcentrez&rendertype=abstract

Behnsen J, Perez-Lopez A, Nuccio SP, Raffatellu M (2015) Exploiting host immunity: the *Salmonella* paradigm. Trends Immunol 36(2):112–120. doi:10.1016/j.it.2014.12.003. http://linkinghub.elsevier.com/retrieve/pii/S1471490614002282

Berger AK, Danthi P (2013) Reovirus activates a caspase-independent cell death pathway. mBio 4 (3):e00178–e001713. doi:10.1128/mBio.00178-13. http://www.pubmedcentral.nih.gov/articlerender.fcgi?artid=3656442&tool=pmcentrez&rendertype=abstract

Berger SB, Kasparcova V, Hoffman S, Swift B, Dare L, Schaeffer M, Capriotti C et al (2014) Cutting edge: RIP1 kinase activity is dispensable for normal development but is a key regulator of inflammation in SHARPIN-deficient mice. J Immunol (Baltimore, Md. : 1950) (May 12). doi:10.4049/jimmunol.1400499. http://www.ncbi.nlm.nih.gov/pubmed/24821972

Bertin J, Armstron RC, Ottilie S, Martin DA, Wang Y, Banks S, Wang GH et al (1997) Death effector domain-containing herpesvirus and poxvirus proteins inhibit both Fas- and TNFR1-induced apoptosis. Proc Natl Acad Sci 94(February):1172–1176

Best SM (2008) NIH public access pp 171–192. doi:10.1146/annurev.micro.62.081307.163009. Viral

Birnbaum MJ, Rollie JC, Lois KM (1994) An apoptosis-inhibiting gene from a nuclear polyhedrosis virus encoding a polypeptide with Cys/His sequence motifs 68(4):2521–2528

Bischofberger M, Lacovache L, Gisou van der Goot F (2012) Pathogenic pore-forming proteins: function and host response. Cell Host Microbe 12(3):266–275. doi:10.1016/j.chom.2012.08.005. http://www.ncbi.nlm.nih.gov/pubmed/22980324

Bliska JB (2003) The ability to replicate in macrophages is conserved between yersinia pestis and *Yersinia pseudotuberculosis* 71(10):5892–5899. doi:10.1128/IAI.71.10.5892

Bonnet MC, Preukschat D, Welz PS, van Loo G, Ermolaeva MA, Bloch W, Haase I, Pasparakis M (2011) The adaptor protein FADD protects epidermal keratinocytes from necroptosis in vivo and prevents skin inflammation. Immunity 35(4):572–582. doi:10.1016/j.immuni.2011.08.014. http://www.ncbi.nlm.nih.gov/pubmed/22000287

Boyden ED, Dietrich WF (2006) Nalp1b controls mouse macrophage susceptibility to anthrax lethal toxin. Nat Genetics 38(2):240–244. doi:10.1038/ng1724. http://www.ncbi.nlm.nih.gov/pubmed/16429160

Cai Z, Jitkaew S, Zhao J, Chiang HC, Choksi S, Liu J, Ward Y, Wu LG, Liu ZG (2014) Plasma membrane translocation of trimerized MLKL protein is required for TNF-induced necroptosis. Nat Cell Biology 16(1):55–65. doi:10.1038/ncb2883. http://www.ncbi.nlm.nih.gov/pubmed/24316671

Cho Y, Challa S, Moquin D, Genga R, Dutta T, Guildford M, Chan MK (2010) NIH public access 137(6):1112–1123. doi:10.1016/j.cell.2009.05.037.Phosphorylation-Driven

Christofferson DE, Li Y, Hitomi J, Zhou W, Upperman C, Zhu H, Gerber SA, Gygi S, Yuan J (2012) A novel role for RIP1 kinase in mediating TNF-α production. Cell Death Dis 3:e320. doi:10.1038/cddis.2012.64. http://www.pubmedcentral.nih.gov/articlerender.fcgi?artid=3388236&tool=pmcentrez&rendertype=abstract

Christofferson DE, Li Y, Yuan J (2014) Control of life-or-death decisions by RIP1 kinase. Ann Rev Physiol 76:129–150. doi:10.1146/annurev-physiol-021113-170259. http://www.ncbi.nlm.nih.gov/pubmed/24079414

Christofferson DE, Yuan J (2010) Necroptosis as an alternative form of programmed cell death. Current Opinion Cell Biol 22(2):263–268. doi:10.1016/j.ceb.2009.12.003. http://www.pubmedcentral.nih.gov/articlerender.fcgi?artid=2854308&tool=pmcentrez&rendertype=abstract

Clarke P, Meintzer SM, Moffitt LA, Tyler KL (2003) Two distinct phases of virus-induced nuclear factor kappa B regulation enhance tumor necrosis factor-related apoptosis-inducing ligand-mediated apoptosis in virus-infected cells. J Biol Chem 278(20):18092–18100. doi:10.1074/jbc.M300265200. http://www.ncbi.nlm.nih.gov/pubmed/12637521

Craven RR, Gao X, Allen IC, Gris D, Bubeck Wardenburg J, McElvania-Tekippe E, Ting JP, Duncan JA (2009) *Staphylococcus aureus* alpha-hemolysin activates the NLRP3-inflammasome in human and mouse monocytic cells. PloS One 4(10):e7446. doi:10.1371/journal.pone.0007446. http://www.pubmedcentral.nih.gov/articlerender.fcgi?artid=2758589&tool=pmcentrez&rendertype=abstract

Crook NE, Clem RJ, Miller LK (1993) An apoptosis-inhibiting baculovirus gene with a zinc finger-like motif 67(4):2168–2174

Cuda CM, Misharin AV, Gierut AK, Saber R, Haines GK, Hutcheson J, Hedrick SM et al (2014) Caspase-8 acts as a molecular rheostat to limit RIPK1- and MyD88-mediated dendritic cell activation. J Immunol (Baltimore, Md. : 1950). doi:10.4049/jimmunol.1400122. http://www.ncbi.nlm.nih.gov/pubmed/24808358

Cusson-Hermance N, Khurana S, Lee TH, Fitzgerald KA, Kelliher MA (2005) Rip1 mediates the trif-dependent Toll-like receptor 3- and 4-induced NF-{kappa}B activation but does not contribute to interferon regulatory factor 3 activation. J Biol Chem 280(44):36560–36566. doi:10.1074/jbc.M506831200. http://www.ncbi.nlm.nih.gov/pubmed/16115877

Dannappel M, Vlantis K, Kumari S, Polykratis A, Kim C, Wachsmuth L, Eftychi C et al (2014) RIPK1 maintains epithelial homeostasis by inhibiting apoptosis and necroptosis. Nature 513 (7516):90–94. doi:10.1038/nature13608. http://dx.doi.org/10.1038/nature13608

Danthi P, Holm GH, Stehle T, Dermody TS (2013) Chapter 3 Reovirus receptors, cell entry, and proapoptotic signaling pp 42–71

Degterev A, Huang Z, Boyce M, Li Y, Jagtap P, Mizushima N, Cuny GD, Mitchison TJ, Moskowitz MA, Yuan J (2005) Chemical inhibitor of nonapoptotic cell death with therapeutic potential for ischemic brain injury. Nat Chem Biol 1(2):112–119. doi:10.1038/nchembio711. http://www.ncbi.nlm.nih.gov/pubmed/16408008

Degterev A, Hitomi J, Germscheid M, Ch'en IL, Korkina O, Teng X, Abbott D et al (2008) Identification of RIP1 kinase as a specific cellular target of necrostatins. Nat Chem Biol 4 (5):313–321. doi:10.1038/nchembio.83. http://www.ncbi.nlm.nih.gov/pubmed/18408713

Degterev A, Maki JL, Yuan J (2013) Activity and specificity of necrostatin-1, small-molecule inhibitor of RIP1 kinase. Cell Death Differ 20(2):366. doi:10.1038/cdd.2012.133. http://www.pubmedcentral.nih.gov/articlerender.fcgi?artid=3554332&tool=pmcentrez&rendertype=abstract

Demmler-Harrison GJ (2015) Congenital cytomegalovirus infection: clinical features and diagnosis. Uptodate. http://www.uptodate.com/contents/congenital-cytomegalovirus-infection-clinical-features-and-diagnosis?source=search_result&search=cmv&selectedTitle=2 ~ 150

Deveraux QL, Reed JC (1999) REVIEW IAP family proteins—suppressors of apoptosis, pp 239–252

Dillon CP, Weinlich R, Rodriguez DA, Cripps JG, Quarato G, Gurung P, Verbist KC et al (2014) RIPK1 blocks early postnatal lethality mediated by caspase-8 and RIPK3. Cell 157(5):1189 1202. doi:10.1016/j.cell.2014.04.018. http://www.ncbi.nlm.nih.gov/pubmed/24813850

Goldstein DJ, Weller SK (1988) Factor(s) present in herpes simplex virus type 1-infected cells can compensate for the loss of the large subunit of the viral ribonucleotide reductase: characterization of an ICP6 deletion mutant. Virology 166(1):41–51

Dobbelstein M, Shenk T, Hughes H (1996) Protection against apoptosis by the vaccinia virus SPI-2 (B13R) gene product. Mol Biol 70(9):6479–6485

Dolin R (2015) Epidemiology of influenza. Uptodate. http://www.uptodate.com/contents/epidemiology-of-influenza?source=search_result&search=Dolin+INfluenza&selectedTitle=3 ~ 150

Dondelinger Y, Aguileta MA, Goossens V, Dubuisson C, Grootjans S, Dejardin E, Vandenabeele P, Bertrand MJM (2013) RIPK3 contributes to TNFR1-mediated RIPK1 kinase-dependent apoptosis in conditions of cIAP1/2 depletion or TAK1 kinase inhibition. Cell Death Differ 20(10):1381–1392. doi:10.1038/cdd.2013.94. http://www.pubmedcentral.nih.gov/articlerender.fcgi?artid=3770330&tool=pmcentrez&rendertype=abstract

Dufour F, Sasseville A, Chabaud S, Massie B, Siegel RM, Langelier Y (2011) The ribonucleotide reductase R1 subunits of herpes simplex virus types 1 and 2 protect cells against TNF-α- and FasL-Induced apoptosis by interacting with caspase-8. Apoptosis Int J Programmed Cell Death 16(3):256–271. doi:10.1007/s10495-010-0560-2. http://www.ncbi.nlm.nih.gov/pubmed/21107701

Elias PM (2007) The skin barrier as an innate immune element. Semin Immunopathol 3(14):3–14

Faherty CS, Maurelli AT (2008) Staying alive: bacterial inhibition of apoptosis during infection. Trends Microbiol 16(4):173–180. doi:10.1016/j.tim.2008.02.001. http://www.pubmedcentral.nih.gov/articlerender.fcgi?artid=2746948&tool=pmcentrez&rendertype=abstract

Friel TJ (2014) Epidemiology, clinical manifestations, and treatment of cytomegalovirus infection in immunocompetent adults. Uptodate. http://www.uptodate.com/contents/epidemiology-clinical-manifestations-and-treatment-of-cytomegalovirus-infection-in-immunocompetent-adults?source=search_result&search=cmv&selectedTitle=1 ~ 150

Galluzzi L, Kepp O, Morselli E, Vitale I, Senovilla L, Pinti M, Zitvogel L, Kroemer G (2010) Viral strategies for the evasion of immunogenic cell death. J Intern Med 267(5):526–542. doi:10.1111/j.1365-2796.2010.02223.x. http://www.ncbi.nlm.nih.gov/pubmed/20433579

Günther C, Martini E, Wittkopf N, Amann K, Weigmann B, Neumann H, Waldner MJ et al (2011) Caspase-8 regulates TNF-A-induced epithelial necroptosis and terminal Ileitis. Nature 477 (7364):335–339. doi:10.1038/nature10400. http://www.pubmedcentral.nih.gov/articlerender.fcgi?artid=3373730&tool=pmcentrez&rendertype=abstract

Guo W, Allan AM, Zong R, Zhang L, Johnson EB, Schaller EG, Murthy AC et al (2011) Ablation of Fmrp in adult neural stem cells disrupts hippocampus-dependent learning. Nat Med 17 (5):559–565. doi:10.1038/nm.2336. http://www.pubmedcentral.nih.gov/articlerender.fcgi?artid=3140952&tool=pmcentrez&rendertype=abstract

Guo H, Omoto S, Harris PA, Finger JN, Bertin J, Gough PJ, Kaiser WJ, Mocarski ES (2015) Herpes simplex virus suppresses necroptosis in human cells. Cell Host Microbe 17(2):243–251. doi:10.1016/j.chom.2015.01.003. http://www.ncbi.nlm.nih.gov/pubmed/25674983

He S, Wang L, Miao L, Wang T, Du F, Zhao L, Wang X (2009) Receptor interacting protein kinase-3 determines cellular necrotic response to TNF-alpha. Cell 137(6):1100–1111. doi:10.1016/j.cell.2009.05.021. http://www.ncbi.nlm.nih.gov/pubmed/19524512

Hildebrand JM, Tanzer MC, Lucet IS, Young SN, Spall SK, Sharma P, Pierotti C et al (2014) Activation of the pseudokinase MLKL unleashes the four-helix bundle domain to induce membrane localization and necroptotic cell death. Proc Nat Acad Sci USA 111(42). doi:10.1073/pnas.1408987111. http://www.ncbi.nlm.nih.gov/pubmed/25288762

Hitomi J, Christofferson DE, Ng A, Yao J, Degterev A, Xavier RJ, Yuan J (2008) Identification of a molecular signaling network that regulates a cellular necrotic cell death pathway. Cell 135 (7):1311–1323. doi:10.1016/j.cell.2008.10.044. http://www.pubmedcentral.nih.gov/articlerender.fcgi?artid=2621059&tool=pmcentrez&rendertype=abstract

Hohmann EL (2014) Microbiology and epidemiology of salmonellosis. Uptodate. http://www.uptodate.com/contents/microbiology-and-epidemiology-of-salmonellosis?source=search_result&search=Salmonella&selectedTitle=2 ~ 150

Hu S, Vincenz C, Buller M, Dixit VM (1997) A novel family of viral death effector domain-containing molecules that inhibit both C-95 an tumor necrosis factor receptor-1-induced apoptosis *. J Biol Chem 272(15):9621–9625

Huang Z, Wu SQ, Liang Y, Zhou X, Chen W, Li L, Wu J et al (2015) RIP1/RIP3 binding to HSV-1 ICP6 initiates necroptosis to restrict virus propagation in mice. Cell Host Microbe 17(2):229–242. doi:10.1016/j.chom.2015.01.002. http://www.ncbi.nlm.nih.gov/pubmed/25674982

Kabsch K, Alonso A (2002) The human papillomavirus type 16 E5 protein impairs TRAIL- and FasL-mediated apoptosis in HaCaT cells by different mechanisms. J Virol 76(23):12162–12172. doi:10.1128/JVI.76.23.12162-12172.2002. http://jvi.asm.org/cgi/doi/10.1128/JVI.76.23.12162-12172.2002

Kaczmarek A, Vandenabeele P, Krysko DV (2013) Necroptosis: the release of damage-associated molecular patterns and its physiological relevance. Immunity 38(2):209–223. doi:10.1016/j.immuni.2013.02.003. http://www.ncbi.nlm.nih.gov/pubmed/23438821

Kaiser WJ, Sridharan H, Huang C, Mandal P, Upton JW, Gough PJ, Sehon CA, Marquis RW, Bertin J, Mocarski ES (2013) Toll-like receptor 3-mediated necrosis via TRIF, RIP3, and MLKL. J Biol Chem 288(43):31268–31279. doi:10.1074/jbc.M113.462341. http://www.ncbi.nlm.nih.gov/pubmed/24019532

Kaiser WJ, Daley-Bauer LP, Thapa RJ, Mandal P, Berger SB, Huang C, Sundararajan A et al (2014) RIP1 suppresses innate immune necrotic as well as apoptotic cell death during mammalian parturition. Proc Nat Acad Sci USA 111(21):7753–7758. doi:10.1073/pnas.1401857111. http://www.ncbi.nlm.nih.gov/pubmed/24821786

Kang TB, Yang SH, Toth B, Kovalenko A, Wallach D (2014) Activation of the NLRP3 inflammasome by proteins that signal for necroptosis. Methods Enzymol, 1st ed, vol 545. Elsevier Inc. doi:10.1016/B978-0-12-801430-1.00003-2. http://www.ncbi.nlm.nih.gov/pubmed/25065886

Kaser A, Zeissig S, Blumberg RS (2010) Inflammatory bowel disease. Ann Rev Immunol 28:573–621. doi:10.1146/annurev-immunol-030409-101225. http://www.ncbi.nlm.nih.gov/pubmed/20192811

Kearney CJ, Cullen SP, Clancy D, Martin SJ (2014) RIPK1 can function as an inhibitor rather than an initiator of RIPK3-dependent necroptosis. FEBS J. doi:10.1111/febs.13034. http://www.ncbi.nlm.nih.gov/pubmed/25195660

Kelliher MA, Grimm S, Ishida Y, Kuo F, Stanger BZ, Leder P (1998) The death domain kinase RIP mediates the TNF-induced NF—B signal 8:297–303

Kennedy CL, Smith DJ, Lyras D, Chakravorty A, Rood JI (2009) Programmed cellular necrosis mediated by the pore-forming alpha-toxin from *Clostridium septicum*. PLoS Pathogens 5(7): e1000516. doi:10.1371/journal.ppat.1000516. http://www.pubmedcentral.nih.gov/articlerender.fcgi?artid=2705182&tool=pmcentrez&rendertype=abstract

Kim SJ, Jianrong Li (2013) Caspase blockade induces RIP3-mediated programmed necrosis in Toll-like receptor-activated microglia. Cell Death Dis 4(7):e716. doi:10.1038/cddis.2013.238. http://www.pubmedcentral.nih.gov/articlerender.fcgi?artid=3730412&tool=pmcentrez&rendertype=abstract

Kitur K, Parker D, Nieto P, Ahn DS, Cohen TS, Chung S, Wachtel S, Bueno S, Prince A (2015) Toxin-induced necroptosis is a major mechanism of *Staphylococcus aureus* lung damage. PLoS Pathogens 11(4):e1004820. doi:10.1371/journal.ppat.1004820. http://www.ncbi.nlm.nih.gov/pubmed/25880560

Klein RS (2015a) Clinical manifestations and diagnosis of herpes simplex virus type 1 infection. Uptodate http://www.uptodate.com/contents/clinical-manifestations-and-diagnosis-of-herpes-simplex-virus-type-1-infection?source=search_result&search=HSV&selectedTitle=1 ~ 150

Klein RS (2015b) Herpes simplex virus type 1 encephalitis. Uptodate http://www.uptodate.com/contents/herpes-simplex-virus-type-1-encephalitis?source=search_result&search=HSV+encephalitis&selectedTitle=1 ~ 46

Knapp O, Maier E, Mkaddem SB, Benz R, Bens M, Chenal A, Geny B, Vandewalle A, Popoff MR (2010) *Clostridium septicum* alpha-toxin forms pores and induces rapid cell necrosis. Toxicon Official J Int Soc Toxinol 55(1):61–72. doi:10.1016/j.toxicon.2009.06.037. http://www.ncbi.nlm.nih.gov/pubmed/19632260

Kotton CN, Hohmann EL (2013) Pathogenesis of salmonella gastroenteritis. Uptodate http://www.uptodate.com/contents/pathogenesis-of-salmonella-gastroenteritis?source=search_result&search=Salmonella&selectedTitle=3 ~ 150

Kovalenko A, Kim JC, Kang TB, Rajput A, Bogdanov K, Dittrich-Breiholz O, Kracht M, Brenner O, Wallach D (2009) Caspase-8 deficiency in epidermal keratinocytes triggers an inflammatory skin disease. J Exp Med 206(10):2161–2177. doi:10.1084/jem.20090616. http://www.pubmedcentral.nih.gov/articlerender.fcgi?artid=2757876&tool=pmcentrez&rendertype=abstract

Kumar V, Abul AK, Fausto N, Aster J (2010) Pathologic basis of disease, 8th ed. Saunders

Lamkanfi M, Dixit VM (2010) Manipulation of host cell death pathways during microbial infections. Cell Host Microbe 8(1):44–54. doi:10.1016/j.chom.2010.06.007. http://www.ncbi.nlm.nih.gov/pubmed/20638641

Langelier Y, Bergeron S, Chabaud S, Lippens J, Guilbault C, Sasseville AM, Denis S, Mosser DD, Massie B (2002) The R1 subunit of herpes simplex virus ribonucleotide reductase protects cells against apoptosis at, or upstream of, caspase-8 activation, pp 2779–2789

LaRock CN, Cookson BT (2012) The yersinia virulence effector YopM binds caspase-1 to arrest inflammasome assembly and processing. Cell Host Microbe 12(6):799–805. doi:10.1016/j.chom.2012.10.020. http://www.pubmedcentral.nih.gov/articlerender.fcgi?artid=3703949&tool=pmcentrez&rendertype=abstract

Li M, Beg AA (2000) Induction of necrotic-like cell death by tumor necrosis factor alpha and caspase inhibitors : novel mechanism for killing virus-infected cells 74(16):7470–7477

Li J, McQuade T, Siemer AB, Napetschnig J, Moriwaki K, Hsiao YS, Damko E et al (2012) The RIP1/RIP3 necrosome forms a functional amyloid signaling complex required for programmed necrosis. Cell 150(2):339–350. doi:10.1016/j.cell.2012.06.019. http://www.pubmedcentral.nih.gov/articlerender.fcgi?artid=3664196&tool=pmcentrez&rendertype=abstract

Li S, Zhang L, Yao Q, Li L, Dong N, Rong J, Gao W et al (2013) Pathogen blocks host death-receptor signalling by arginine GlcNAcylation of death domains. Nature 501(7466):242–246. doi:10.1038/nature12436. http://www.ncbi.nlm.nih.gov/pubmed/23955153

Li L, Chen W, Liang Y, Ma H, Li W, Zhou Z, Li J et al (2014) The Gβγ-src signaling pathway regulates TNF-induced necroptosis via control of necrosome translocation. Cell Res 2:1–16. doi:10.1038/cr.2014.17. http://www.ncbi.nlm.nih.gov/pubmed/24513853

Lin CF, Chen CL, Huang WC, Cheng YL, Hsieh CY, Wang CY, Hong MY (2010) Different types of cell death induced by enterotoxins. Toxins 2(8):2158–2176. doi:10.3390/toxins2082158. http://www.pubmedcentral.nih.gov/articlerender.fcgi?artid=3153280&tool=pmcentrez&rendertype=abstract

Linkermann A, Green DR (2014) Necroptosis. New Engl J Med 370(5):455–465. doi:10.1056/NEJMra1310050. http://www.pubmedcentral.nih.gov/articlerender.fcgi?artid=4035222&tool=pmcentrez&rendertype=abstract

Lukens JR, Vogel P, Johnson GR, Kelliher MA, Iwakura Y, Lamkanfi M, Kanneganti TD (2013) RIP1-driven autoinflammation targets IL-1α independently of inflammasomes and RIP3. Nature 498(7453):224–227. doi:10.1038/nature12174. http://www.pubmedcentral.nih.gov/articlerender.fcgi?artid=3683390&tool=pmcentrez&rendertype=abstract

Macdonald TT, Monteleone G (2005) Immunity, inflammation, and allergy in the gut pp 1920–1925

Macen JL, Garner RS, Musyt PY, Brookst MA, Turnert PC, Moyert RW, Mcfadden G, Bleackley RC (1996) Orthopoxvirus cytokine 93:9108–9113

McComb S, Cheung HH, Korneluk RG, Wang S, Krishnan L, Sad S (2012) cIAP1 and cIAP2 limit macrophage necroptosis by inhibiting Rip1 and Rip3 activation. Cell Death Differ 19 (11):1791–1801. doi:10.1038/cdd.2012.59. http://www.pubmedcentral.nih.gov/articlerender.fcgi?artid=3469059&tool=pmcentrez&rendertype=abstract

McCoy MW, Moreland SM, Detweiler CS (2012) Hemophagocytic macrophages in murine typhoid fever have an anti-inflammatory phenotype. Infect Immun 80(10):3642–3649. doi:10.1128/IAI.00656-12. http://www.pubmedcentral.nih.gov/articlerender.fcgi?artid=3457584&tool=pmcentrez&rendertype=abstract

Meylan E, Burns K, Hofmann K, Blancheteau V, Martinon F, Kelliher M, Tschopp J (2004) RIP1 is an essential mediator of Toll-like receptor 3-induced NF-kappa B activation. Nat Immunol 5 (5):503–507. doi:10.1038/ni1061. http://www.ncbi.nlm.nih.gov/pubmed/15064760

Miller EA, Ernst JD (2009) Anti-TNF immunotherapy and tuberculosis reactivation: another mechanism revealed. J Clin Invest 119(5):1079–1082. doi:10.1172/JCI39143. http://www.jci.org/articles/view/39143

Mocarski ES, Upton JW, Kaiser WJ (2012) Viral infection and the evolution of caspase 8-regulated apoptotic and necrotic death pathways. Nat Rev Immunol 12(2):79–88. doi:10.1038/nri3131. http://www.ncbi.nlm.nih.gov/pubmed/22193709

Moquin DM, McQuade T, Chan FK (2013) CYLD deubiquitinates RIP1 in the TNF-α-induced necrosome to facilitate kinase activation and programmed necrosis. PloS One 8(10):e76841. doi:10.1371/journal.pone.0076841. http://www.pubmedcentral.nih.gov/articlerender.fcgi?artid=3788787&tool=pmcentrez&rendertype=abstract

Moriwaki K, Balaji S, McQuade T, Malhotra N, Kang J, Chan FK (2014) The necroptosis adaptor RIPK3 promotes injury-induced cytokine expression and tissue repair. Immunity 41(4):567–578. doi:10.1016/j.immuni.2014.09.016. http://linkinghub.elsevier.com/retrieve/pii/S1074761314003501

Moriwaki K, Bertin J, Gough PJ, Chan FK (2015) A RIPK3-caspase 8 complex mediates atypical pro-IL-1β processing. J Immunol (Baltimore, Md. : 1950) 194(4):1938–1944. doi:10.4049/jimmunol.1402167. http://www.ncbi.nlm.nih.gov/pubmed/25567679

Murphy JM, Czabotar PE, Hildebrand JM, Lucet IS, ZhangJG, Alvarez-Diaz S, Lewis R et al (2013) The pseudokinase MLKL mediates necroptosis via a molecular switch mechanism. Immunity 39(3):443–453. doi:10.1016/j.immuni.2013.06.018. http://www.ncbi.nlm.nih.gov/pubmed/24012422

Nestle FO, Meglio PD, Qin JZ, Nickoloff BJ (2009) Skin immune sentinels in health and disease. Nat Rev Immunol 9(10):679–691. doi:10.1038/nri2622. http://www.pubmedcentral.nih.gov/articlerender.fcgi?artid=2947825&tool=pmcentrez&rendertype=abstract

Nix RN, Altschuler SE, Henson PM, Detweiler CS (2007) Hemophagocytic macrophages harbor salmonella enterica during persistent infection. PLoS Pathog 3(12):e193. doi:10.1371/journal.ppat.0030193. http://www.pubmedcentral.nih.gov/articlerender.fcgi?artid=2134957&tool=pmcentrez&rendertype=abstract

O'Ryan MG., Matson DO (2015) Clinical manifestations and diagnosis of rotavirus infection. Uptodate http://www.uptodate.com/contents/clinical-manifestations-and-diagnosis-of-rotavirus-infection?source=search_result&search=rotavirus&selectedTitle=1 ~ 82

Ofengeim D, Yuan J (2013) Regulation of RIP1 kinase signalling at the crossroads of inflammation and cell death. Nat Rev Mol Cell Biol 14(11):727–736. doi:10.1038/nrm3683. http://www.ncbi.nlm.nih.gov/pubmed/24129419

Ofengeim D, Ito Y, Najafov A, Zhang Y, Shan B, DeWitt JP, Ye J et al (2015) Activation of necroptosis in multiple sclerosis. Cell Rep 10(11):1836–1849. doi:10.1016/j.celrep.2015.02.051. http://www.ncbi.nlm.nih.gov/pubmed/25801023

Omoto S, Guo H, Talekar GR, Roback L, Kaiser WJ, Mocarski E (2015) Suppression of RIP3-dependent necroptosis by human cytomegalovirus. J Biol Chem (404). doi:10.1074/jbc.M115.646042. http://www.ncbi.nlm.nih.gov/pubmed/25778401

Orozco S, Yatim N, Werner MR, Tran H, Gunja SY, Wg Tait S, Albert ML, Green DR, Oberst A (2014) RIPK1 both positively and negatively regulates RIPK3 oligomerization and necroptosis. Cell Death Differ 21(10):1511–1521. doi:10.1038/cdd.2014.76. http://www.ncbi.nlm.nih.gov/pubmed/24902904

Osborn SL, Diehl G, Han SJ, Xue L, Kurd N, Hsieh K, Cado D, Robey EA, Winoto A (2010) Fas-associated death domain (FADD) is a negative regulator of T-cell receptor-mediated necroptosis. Proc Nat Acad Sci USA 107(29):13034–13039. doi:10.1073/pnas.1005997107. http://www.pubmedcentral.nih.gov/articlerender.fcgi?artid=2919948&tool=pmcentrez&rendertype=abstract

Pearson JS, Giogha C, Ong SY, Kennedy CL, Kelly M, Robinson KS, Fok Lung TW et al (2013) A type III effector antagonizes death-receptor signalling during bacterial gut infection. Nature

501(7466):247–251. doi:10.1038/nature12524. http://www.nature.com/doifinder/10.1038/nature12524

Philip NH, Dillon CP, Snyder AG, Fitzgerald P, Wynosky-Dolfi MA, Zwack EE, Hu B et al (2014) Caspase-8 mediates caspase-1 processing and innate immune defense in response to bacterial blockade of NF-κB and MAPK signaling. Proc Nat Acad Sci USA 111(20):7385–7390. doi:10.1073/pnas.1403252111. http://www.ncbi.nlm.nih.gov/pubmed/24799700

Polykratis A, Hermance N, Zelic M, Roderick J, Kim C, Van TM, Lee TH, Chan FKM, Pasparakis M, Kelliher MA (2014) Cutting edge: RIPK1 kinase inactive mice are viable and protected from TNF-induced necroptosis in vivo. J Immunol (Baltimore, Md. : 1950). doi:10.4049/jimmunol.1400590. http://www.ncbi.nlm.nih.gov/pubmed/25015821

Pozniak A (2015) Clinical manifestations and evaluation of pulmonary tuberculosis. Uptodate http://www.uptodate.com/contents/clinical-manifestations-and-evaluation-of-pulmonary-tuberculosis?source=search_result&search=Clinical+manifestations+and+evaluation+of+pulmonary+tb&selectedTitle=1 ~ 107

Pujol C, Bliska JB (2003) The ability to replicate in macrophages is conserved between yersinia pestis and yersinia pseudotuberculosis. Infect Immun 71(10):5892–5899. doi:10.1128/IAI.71.10.5892-5899.2003. http://iai.asm.org/cgi/doi/10.1128/IAI.71.10.5892-5899.2003

Pujol C, Bliska JB (2005) Turning yersinia pathogenesis outside in: subversion of macrophage function by intracellular yersiniae. Clin Immunol (Orlando, Fla.) 114(3):216–226. doi:10.1016/j.clim.2004.07.013. http://www.ncbi.nlm.nih.gov/pubmed/15721832

Pujol C, Klein KA, Romanov GA, Palmer LE, Cirota C, Zhao Z, Bliska JB (2009) Yersinia pestis can reside in autophagosomes and avoid xenophagy in murine macrophages by preventing vacuole acidification. Infect Immun 77(6):2251–2261. doi:10.1128/IAI.00068-09. http://www.pubmedcentral.nih.gov/articlerender.fcgi?artid=2687347&tool=pmcentrez&rendertype=abstract

Rajput A, Kovalenko A, Bogdanov K, Yang SH, Kang TB, Kim JC, Du J, Wallach D (2011) RIG-I RNA helicase activation of IRF3 transcription factor is negatively regulated by caspase-8-mediated cleavage of the RIP1 protein. Immunity 34(3):340–351. doi:10.1016/j.immuni.2010.12.018. http://www.ncbi.nlm.nih.gov/pubmed/21419663

Richardson-Burns SM, Kominsky DJ, Tyler KL (2002) Reovirus-induced neuronal apoptosis is mediated by caspase 3 and is associated with the activation of death-receptors, pp 365–380. doi:10.1080/13550280290100806

Rickard JA, O'Donnell JA, Evans JM, Lalaoui N, Poh AR, Rogers T, Vince JE et al (2014) RIPK1 regulates RIPK3-MLKL-driven systemic inflammation and emergency hematopoiesis. Cell 8:1–14. doi:10.1016/j.cell.2014.04.019. http://www.ncbi.nlm.nih.gov/pubmed/24813849

Robinson N, McComb S, Mulligan R, Dudani R, Krishnan L, Sad S (2012) Type i interferon induces necroptosis in macrophages during infection with *Salmonella enterica* serovar typhimurium. Nat Immunol 13(10):954–962. doi:10.1038/ni.2397. http://www.pubmedcentral.nih.gov/articlerender.fcgi?artid=4005791&tool=pmcentrez&rendertype=abstract

Roca FJ, Ramakrishnan L (2013) TNF dually mediates resistance and susceptibility to mycobacteria via mitochondrial reactive oxygen species. Cell 153(3):521–534. doi:10.1016/j.cell.2013.03.022. http://www.pubmedcentral.nih.gov/articlerender.fcgi?artid=3790588&tool=pmcentrez&rendertype=abstract

Rodrigue-Gervais IG, Labbé K, Dagenais M, Dupaul-Chicoine J, Champagne C, Morizot A, Skeldon A et al (2014) Cellular inhibitor of apoptosis protein cIAP2 protects against pulmonary tissue necrosis during influenza virus infection to promote host survival. Cell Host Microbe 15(1):23–35. doi:10.1016/j.chom.2013.12.003. http://www.ncbi.nlm.nih.gov/pubmed/24439895

Roy CR, Mocarski ES (2007) Pathogen subversion of cell-intrinsic innate immunity. Nat Immunol 8(11):1179–1187. doi:10.1038/ni1528. http://www.ncbi.nlm.nih.gov/pubmed/17952043

Schluger NW, Rom WN (1998) The host immune response to tuberculosis. Am J Respir Crit Care Med 157(19). Retrieved from http://www.atsjournals.org/doi/abs/10.1164/ajrccm.157.3.9708002

Sexton DJ (2014) Epidemiology, microbiology and pathogenesis of plague (yersinia pestis infection). Uptodate http://www.uptodate.com/contents/epidemiology-microbiology-and-pathogenesis-of-plague-yersinia-pestis-infection?source=search_result&search=Pestis&selectedTitle=1 ~ 25

Shisler JL, Moss B (2001) Molluscum contagiosum virus inhibitors of apoptosis: the MC159 v-FLIP protein blocks Fas-induced activation of procaspases and degradation of the related MC160 protein. Virology 282(1):14–25. doi:10.1006/viro.2001.0834. http://www.ncbi.nlm.nih.gov/pubmed/11259186

Skaletskaya A, Bartle LM, Chittenden T, McCormick AL, Mocarski ES, Goldmacher VS (2001) A cytomegalovirus-encoded inhibitor of apoptosis that suppresses caspase-8 activation. Proc Nat Acad Sci USA 98(14):7829–7834. doi:10.1073/pnas.141108798. http://www.pubmedcentral.nih.gov/articlerender.fcgi?artid=35427&tool=pmcentrez&rendertype=abstract

Sridharan H, Upton JW (2014) Programmed necrosis in microbial pathogenesis. Trends Microbiol 22 (4):199–1207. doi:10.1016/j.tim.2014.01.005. http://www.ncbi.nlm.nih.gov/pubmed/24565922

Takahashi N, Vereecke L, Bertrand MJ, Duprez L, Berger SB, Divert T, Gonçalves A et al (2014) RIPK1 ensures intestinal homeostasis by protecting the epithelium against apoptosis. Nature 513(7516):95–99. doi:10.1038/nature13706. http://www.nature.com/doifinder/10.1038/nature13706

Tauxe RV (2013) Clinical manifestations and diagnosis of yersinia infections. Uptodate http://www.uptodate.com/contents/clinical-manifestations-and-diagnosis-of-yersinia-infections?source=search_result&search=Yersinia&selectedTitle=1 ~ 92

Thome M, Schneider P, Hofman K, Ficjenscher H, Meini E, Neipel F, Mattmann C et al (1997) Viral FLICE-inhibitory proteins (FLIPs) prevent apoptosis induced by death-receptors.pdf. Nature 386:517–521

Thurau M, Everett H, Tapernoux M, Tschopp J, Thome M (2006) The TRAF3-binding site of human molluscipox virus FLIP molecule MC159 is critical for its capacity to inhibit Fas-induced apoptosis. Cell Death Differ 13(9):1577–1585. doi:10.1038/sj.cdd.4401847. http://www.ncbi.nlm.nih.gov/pubmed/16410799

Ting AT, Pimentel-Muiffios FX (1996) RIP activation apoptosis 15(22):6189–6196

Tobin DM, Vary JC, Ray JP, Walsh GS, Dunstan SJ, Bang ND, Hagge DA et al (2010) The lta4 h locus modulates susceptibility to mycobacterial infection in zebrafish and humans. Cell 140 (5):717–730. doi:10.1016/j.cell.2010.02.013. http://www.pubmedcentral.nih.gov/articlerender.fcgi?artid=2907082&tool=pmcentrez&rendertype=abstract

Tobin DM, Roca FJ, Oh SF, McFarland R, Vickery TW, Ray JP, Ko DC et al (2012) Host genotype-specific therapies can optimize the inflammatory response to mycobacterial infections. Cell 148(3):434–446. doi:10.1016/j.cell.2011.12.023. http://www.pubmedcentral.nih.gov/articlerender.fcgi?artid=3433720&tool=pmcentrez&rendertype=abstract

Turner SJ, Silke J, Kenshole B, Ruby J (2000) Characterization of the ectromelia virus serpin, SPI-2 64:2425–2430

Ulrichs T, Kaufmann SHE (2006) New insights into the function of granulomas in human tuberculosis. J Pathol 208(2):261–269. doi:10.1002/path.1906. http://www.ncbi.nlm.nih.gov/pubmed/16362982

Upton JW, Kaiser WJ, Mocarski ES (2010) Virus inhibition of RIP3-dependent necrosis. Cell Host Microbe 7(4):302–313. doi:10.1016/j.chom.2010.03.006. http://www.ncbi.nlm.nih.gov/pubmed/20413098

Upton JW, Kaiser WJ, Mocarski ES (2012) DAI/ZBP1/DLM-1 complexes with RIP3 to mediate virus-induced programmed necrosis that is targeted by murine cytomegalovirus vIRA. Cell Host Microbe 11(3):290–297. doi:10.1016/j.chom.2012.01.016. http://www.pubmedcentral.nih.gov/articlerender.fcgi?artid=3531981&tool=pmcentrez&rendertype=abstract

Vanden Berghe T, Linkermann A, Jouan-Lanhouet S, Walczak H, Vandenabeele P (2014) Regulated necrosis: the expanding network of non-apoptotic cell death pathways. Nat Rev Mol Cell Biol 15(2):135–147. doi:10.1038/nrm3737. http://www.ncbi.nlm.nih.gov/pubmed/24452471

Vandenabeele P, Declercq W, Van Herreweghe F, Berghe TV (2010a) The role of the kinases RIP1 and RIP3 in TNF-induced necrosis. Sci Signaling 3(115):re4. doi:10.1126/scisignal.3115re4. http://www.ncbi.nlm.nih.gov/pubmed/20354226

Vandenabeele P, Galluzzi L, Berghe TV, Kroemer G (2010b) Molecular mechanisms of necroptosis: an ordered cellular explosion. Nat Rev Mol Cell Biol 11(10):700–714. doi:10.1038/nrm2970. http://www.ncbi.nlm.nih.gov/pubmed/20823910

Viboud GI, Bliska JB (2005) Yersinia outer proteins: role in modulation of host cell signaling responses and pathogenesis. Ann Rev Microbiol 59:69–89. doi:10.1146/annurev.micro.59.030804.121320. http://www.ncbi.nlm.nih.gov/pubmed/15847602

Wang X, Jiang W, Yan Y, Gong T, Han J, Tian Z, Zhou R (2014a) RNA viruses promote activation of the NLRP3 inflammasome through a RIP1-RIP3-DRP1 signaling pathway. Nat Immunol 15(12). doi:10.1038/ni.3015. http://www.nature.com/doifinder/10.1038/ni.3015

Wang X, Li Y, Liu S, Yu X, Li L, Shi C, He W et al (2014b) Direct activation of RIP3/MLKL-dependent necrosis by herpes simplex virus 1 (HSV-1) protein ICP6 triggers host antiviral defense. Proc Nat Acad Sci USA 111(43):15438–15443. doi:10.1073/pnas.1412767111. http://www.ncbi.nlm.nih.gov/pubmed/25316792

Wanke CA (2013) Pathogenic *Escherichia coli*. Uptodate http://www.uptodate.com/contents/pathogenic-escherichia-coli?source=search_result&search=E.coli&selectedTitle=1 ~ 150

Weinlich R, Oberst A, Dillon CP, Janke LJ, Milasta S, Lukens JR, Rodriguez DA et al (2013) Protective roles for caspase-8 and cFLIP in adult homeostasis. Cell Reports 5(2):340–348. doi:10.1016/j.celrep.2013.08.045. http://www.pubmedcentral.nih.gov/articlerender.fcgi?artid=3843376&tool=pmcentrez&rendertype=abstract

Welz PS, Wullaert A, Vlantis K, Kondylis V, Fernández-Majada V, Ermolaeva M, Kirsch P, Sterner-Kock A, van Loo G, Pasparakis M (2011) FADD prevents RIP3-mediated epithelial cell necrosis and chronic intestinal inflammation. Nature 477(7364):330–334. doi:10.1038/nature10273. http://www.ncbi.nlm.nih.gov/pubmed/21804564

Weng D, Marty-Roix R, Ganesan S, Proulx MK, Vladimer GI, Kaiser WJ, Mocarski ES, et al (2014) Caspase-8 and RIP kinases regulate bacteria-induced innate immune responses and cell death. Proc Nat Acad Sci USA 111(20):7391–7396. doi:10.1073/pnas.1403477111. http://www.ncbi.nlm.nih.gov/pubmed/24799678

Wu J, Huang Z, Ren J, Zhang Z, He P, Li Y, Ma J et al (2013) Mlkl knockout mice demonstrate the indispensable role of mlkl in necroptosis. Cell Res 23(8):994–1006. doi:10.1038/cr.2013.91. http://www.pubmedcentral.nih.gov/articlerender.fcgi?artid=3731568&tool=pmcentrez&rendertype=abstract

Wu XN, Yang ZH, Wang XK, Zhang Y, Wan H, Song Y, Chen X, Shao J, Han J (2014) Distinct roles of RIP1-RIP3 hetero- and RIP3-RIP3 homo-interaction in mediating necroptosis. Cell Death Differ 1–12. doi:10.1038/cdd.2014.77. http://www.ncbi.nlm.nih.gov/pubmed/24902902

Xie T, Peng W, Liu Y, Yan C, Maki J, Degterev A, Yuan J, Shi Y (2013a) Structural basis of RIP1 inhibition by necrostatins. Structure (London, England : 1993) 21(3):493–499. doi:10.1016/j.str.2013.01.016. http://www.ncbi.nlm.nih.gov/pubmed/23473668

Xie T, Peng W, Yan C, Wu J, Gong X, Shi Y (2013b) Structural insights into RIP3-mediated necroptotic signaling. Cell Rep 5(1):70–78. doi:10.1016/j.celrep.2013.08.044. http://www.ncbi.nlm.nih.gov/pubmed/24095729

Yatim N, Albert ML (2011) Dying to replicate: the orchestration of the viral life cycle, cell death pathways, and immunity. Immunity 35(4):478–490. doi:10.1016/j.immuni.2011.10.010. http://www.ncbi.nlm.nih.gov/pubmed/22035840

You Z, Savitz SI, Yang J, Degterev A, Yuan J, Cuny GD, Moskowitz MA, WhalenMJ (2008) Necrostatin-1 reduces histopathology and improves functional outcome after controlled cortical impact in mice. J Cereb Blood Flow Metab Official J Int Soc Cereb Blood Flow Metab 28(9):1564–1573. doi:10.1038/jcbfm.2008.44. http://www.pubmedcentral.nih.gov/articlerender.fcgi?artid=2831087&tool=pmcentrez&rendertype=abstract

Zhang SQ, Kovalenko A, Cantarella G, Wallach D (2000) Recruitment of the IKK signalosome to the p55 TNF receptor: RIP and A20 bind to NEMO (IKKγ) upon receptor stimulation 12:301–311

Zhang DW, Shao J, Lin J, Zhang N, Lu BL, Lin SC, Dong MQ, Han J (2009) RIP3, an energy metabolism regulator that switches TNF-induced cell death from apoptosis to necrosis. Science (New York) 325(5938):332–336. doi:10.1126/science.1172308. http://www.ncbi.nlm.nih.gov/pubmed/19498109

Zhang Y, Romanov G, Bliska JB (2011) Type III secretion system-dependent translocation of ectopically expressed yop effectors into macrophages by intracellular yersinia pseudotuberculosis. Infect Immun 79(11):4322–4331. doi:10.1128/IAI.05396-11. http://www.pubmedcentral.nih.gov/articlerender.fcgi?artid=3257923&tool=pmcentrez&rendertype=abstract

Zhao J, Jitkaew S, Cai Z, Choksi S, Li Q, Luo J, Liu ZG (2012) Mixed lineage kinase domain-like is a key receptor interacting protein 3 downstream component of TNF-induced necrosis. Proc Nat Acad Sci USA 109(14):5322–5327. doi:10.1073/pnas.1200012109. http://www.pubmedcentral.nih.gov/articlerender.fcgi?artid=3325682&tool=pmcentrez&rendertype=abstract

Epithelial Cell Death and Inflammation in Skin

Snehlata Kumari and Manolis Pasparakis

Abstract The presence of dying cells in inflamed tissues has been recognized since many years, but until recently cell death was considered primarily a consequence of inflammation. Recent data in mouse models suggest that cell death could provide a potent trigger of inflammation. The identification of necroptosis as a new type of regulated necrotic cell death that is induced by death receptors, toll like receptors and type I interferon receptor indicated that necroptosis could contribute to the proinflammatory properties of these receptors. This is particularly relevant to the skin, a tissue that provides a life-sustaining structural and immunological barrier with the environment and is constantly exposed to mechanical, chemical, and microbial insults. Studies in mouse models showed that sensitization of keratinocytes to apoptosis or necroptosis triggered by TNF and other stimuli causes severe chronic inflammatory skin lesions. In addition, keratinocyte death is a prominent histopathological feature of many inflammatory skin diseases, suggesting that death of epithelial cells could contribute to the pathogenesis of skin inflammation. Here we review recent studies in genetic mouse models providing evidence that keratinocyte death is a potent trigger of skin inflammation and discuss their potential relevance for human inflammatory skin diseases.

S. Kumari · M. Pasparakis (✉)
Institute for Genetics, University of Cologne, 50674 Cologne, Germany
e-mail: pasparakis@uni-koeln.de

S. Kumari · M. Pasparakis
Cologne Excellence Cluster on Cellular Stress Responses in Aging-Associated Diseases (CECAD), University of Cologne, 50931 Cologne, Germany

S. Kumari · M. Pasparakis
Center for Molecular Medicine (CMMC), University of Cologne, 50931 Cologne, Germany

Current Topics in Microbiology and Immunology (2017) 403:77–94
DOI 10.1007/82_2015_466

Published Online: 12 August 2015

Contents

1 Introduction

1.1 The Skin Epithelium

The skin is the largest organ of the body and provides the organism with a life-sustaining interface with the environment. The skin is composed of two compartments, the epidermis and the dermis. The epidermis constitutes a stratified epithelium composed of keratinocytes and small numbers of Langerhans cells, γδ-T cells and Merkel cells. Keratinocytes are organized in structured layers forming a mechanical barrier with the outside environment, which prevents water loss and protects the organism from mechanical, chemical, and microbial challenges. Keratinocytes in the basal layer proliferate and move upwards to the suprabasal layer where they stop proliferating and start differentiating. Terminally differentiated keratinocytes form the upper cornified layer of the epidermis. During this cornification process, keratinocytes become metabolically inactive, lose their organelles and undergo a regulated process of cell death that is up to now poorly understood. Under physiological conditions, the skin does not show cell death-associated markers, such as DNA fragmentation or activation of caspases (Lippens et al. 2000, 2005). The dermis primarily consists of fibroblasts and is rich in extracellular matrix, blood vessels, and nerve endings. The dermis also contains different types of immune cells including T and B lymphocytes, macrophages, neutrophils, dendritic cells, and mast cells (Nestle et al. 2009; Pasparakis et al. 2014).

Skin homeostasis is maintained by the complex communication between the different types of cells present in the epidermis and the dermis. In addition to their key function in forming a mechanical barrier, keratinocytes are now recognized for their important role in regulating immune responses in the skin (Di Meglio et al. 2011).

Epidermal keratinocytes express receptors that sense microbes but also danger signals and trigger the production of soluble mediators such as cytokines and chemokines that communicate with immune cells to coordinate immune responses in the skin. Over the last years, several studies using genetically modified mouse models provided evidence that keratinocytes regulate skin inflammation. Deregulation of keratinocyte responses to stress induced by cytokines, microbes, or other danger signals often resulted in disruption of skin homeostasis and the pathogenesis of inflammatory skin conditions resembling human chronic skin diseases such as psoriasis. Altered expression of cytokines and also the sensitization of keratinocytes to different types of cell death were identified as potent triggers of skin inflammation in several mouse models (Pasparakis et al. 2014).

1.2 NF-κB Signaling

The nuclear factor kappa B (NF-κB) family of transcription factor consists of five members in mammals, p65 (RelA), RelB, c-Rel, p50/105 (NF-κB1), and p52/100 (NF-κB2), which form homo- and heterodimers (Hayden and Ghosh 2008). NF-κB dimers remain inactive by association with inhibitory proteins of the inhibitor of NF-κB (IκB) family, consisting of IκBα, IκBβ, IκBγ, IκBε, BCL-3, and the precursor proteins p100 and p105 (Hayden and Ghosh 2008). Cell activation by a large number of stimuli results in the activation of the IκB kinase (IKK) complex, consisting of the catalytic IKK1/IKKα and IKK2/IKKβ subunits and the regulator NEMO/IKKγ subunit, which phosphorylates IκB proteins targeting them for polyubiquitination and proteasome-dependent degradation. This allows NF-κB dimers to accumulate in the nucleus and activate the transcription of a large number of target genes including cytokines, chemokines, adhesion molecules and proteins regulating cell proliferation, survival and death (Hayden and Ghosh 2008). The NF-κB-dependent expression of NF-κB inhibitors such as A20, CYLD, and IκB provides negative feedback control limiting NF-κB activation (Chen et al. 2006; Hayden and Ghosh 2008).

1.3 Regulated Cell Death: Apoptosis and Necroptosis

Regulated cell death was for many years synonymous to apoptosis, while necrosis was considered a form of accidental cell death that is not molecularly controlled. Apoptosis is morphologically characterized by chromatin condensation, nuclear fragmentation, blebbing of the plasma membrane and formation of apoptotic bodies. Apoptosis is molecularly controlled by a family of proteases called caspases. The extrinsic apoptotic pathway is triggered by death receptors, such as TNFR1, FAS, and TRAILR (Wilson et al. 2009). Death receptor activation results in the formation of a complex between the Fas-associated with death domain protein (FADD) and caspase-8, which facilitates the activation of caspase-8. Activation of caspase-8 subsequently induces the cleavage and activation of caspase-3, which then cleaves a number of cellular proteins resulting in apoptosis. Apoptosis controls several developmental processes in the body both during embryogenesis and in the adult life, such as the removal of vestigial tissues, the formation of limb digits, neuronal development, and lymphocyte development (Jacobson et al. 1997; Milligan and Schwartz 1997).

Necroptosis was identified as a regulated type of necrotic cell death that is morphologically characterized by rapid permeabilization of the plasma membrane, increased cell volume, swelling of organelles, lysosomal membrane permeabilization and mild chromatin condensation (Pasparakis and Vandenabeele 2015). Receptor interacting protein kinase-3 (RIPK3) and mixed lineage kinase domain-like (MLKL) are required for the induction of necroptosis (Cho et al. 2009; He et al. 2009; Zhang et al. 2009; Vandenabeele et al. 2010; Sun et al. 2012; Zhao et al. 2012; Murphy et al. 2013). Necroptosis is triggered by many upstream pathways including death receptors, TLR3 and TLR4, type I interferon receptor (IFNAR1) and viral DNA-mediated activation of the DNA-dependent activator of IFN-regulatory factors (He et al. 2011; Upton et al. 2012; Kaiser et al. 2013; Thapa et al. 2013; McComb et al. 2014). The RIP homotypic interaction or motif (RHIM)-dependent interaction of RIPK1, RIPK3, TRIF, and DAI is crucial for the induction of necroptosis. Most of the understanding and insights into the molecular mechanisms regulating necroptosis are derived from the study of TNFR1 signaling.

1.4 TNFR1 Signaling

TNFR1 is a potent inducer of pro-inflammatory and pro-survival signaling via the activation of NF-κB and mitogen-activated protein kinase (MAPK) pathways, but also of cell death via both apoptosis and necroptosis. Binding of TNF to TNFR1 triggers the formation of a receptor proximal signaling complex (complex I), composed of TNFR1-associated protein with death domain (TRADD), receptor interacting protein kinase 1 (RIPK1), TNF receptor-associated factor-2 (TRAF2), and inhibitor of apoptosis proteins 1 and 2 (cIAP1 and cIAP2), as well as the linear ubiquitin chain assembly complex (LUBAC) composed of haem-oxidized IRP2 ubiquitin ligase-1 (HOIL-1L), HOIL-1-interacting protein (HOIP) and SHANK-associated RH domain-interacting protein (SHARPIN) (Haas 2009; Haas et al. 2009; Wilson et al. 2009; Gerlach et al. 2011; Ikeda et al. 2011; Tokunaga et al. 2011; Vanden Berghe et al. 2014). Ubiquitination of several proteins in the TNFR1 signaling complex is critical for the activation of downstream signaling. The recruitment of the TAK1/TAB and the IκB kinase (IKK) signaling complexes results in activation of NF-κB and MAPK pathways. Activation of NF-κB is generally believed to prevent TNFR1-induced apoptosis by inducing the expression of anti-apoptotic proteins such as cellular FLICE-like inhibitory proteins (cFLIP) and members of the Bcl family such as Bcl-xl.

When pro-survival signaling is compromised, e.g., upon NF-κB inhibition or blockade of transcription or translation, TNFR1 stimulation triggers the formation of the death-inducing signaling complexes IIa and IIb. Complex IIa contains TRADD, FADD, and caspase-8, while complex IIb contains FADD, caspase-8, RIPK1, RIPK3, and $cFLIP_L$. Complexes IIa and IIb trigger apoptosis via the activation of caspase-8. When caspase-8 is inhibited, a complex consisting of RIPK1, RIPK3, and MLKL, termed the necrosome, forms and triggers necroptosis. Intramolecular autophosphorylation of RIPK3 within the necrosome and the subsequent phosphorylation of MLKL are essential for the induction of necroptosis. Phosphorylated MLKL is recruited to the plasma membrane where it is believed to induce necroptosis by two non-mutually exclusive mechanisms involving pore formation on the lipid bilayer or the regulation of ion channels (Pasparakis and Vandenabeele 2015).

2 Inflammatory and Cell Death Signaling in Skin Homeostasis: Lessons from Mouse Models

Genetic mouse models allowing the analysis of the tissue-specific function of genes and their products have provided invaluable information on the molecular mechanisms and pathways regulating tissue homeostasis and inflammation. In particular in the skin, a number of approaches have been used to address the epidermal keratinocyte-specific function of pathways regulating cell death and inflammation. The availability of promoters such at those of the keratin 14 and keratin 5 genes that are almost exclusively expressed in keratinocytes has facilitated the generation of mice expressing transgenes in keratinocytes. These promoters have been used to express dominant negative or constitutively active molecules but most importantly to express Cre recombinase in keratinocytes. The latter mice have been combined with a large number of mouse models carrying loxP-flanked alleles in order to generate mice with keratinocyte-specific knockout of the respective genes. Here we will discuss a number of studies using keratinocyte-specific gene manipulation to study the role of key components of the pathways regulating cell death and inflammation in keratinocytes.

2.1 *NF-κB Signaling in Keratinocytes Controls Skin Homeostasis and Inflammation*

The important role of NF-κB signaling in skin homeostasis was initially noted in mice lacking IκBα ($I\kappa B\alpha^{-/-}$), which showed over-activation of NF-κB and developed multi-organ inflammation with very prominent involvement of the skin. The manifested skin lesions were characterized by epidermal hyperplasia and infiltration of immune cells and resulting in death of mice within 10 days of birth (Beg et al. 1995; Klement et al. 1996; Rebholz et al. 2007). Genetic deletion of either TNFR1, TNF, LTα or LTβ ameliorated skin inflammation, whereas the combined deficiency of TNF, LTα, and LTβ fully prevented skin lesion development in *IκBα*$^{-/-}$ mice, identifying TNF and lymphotoxins as the key drivers of inflammation (Rebholz et al. 2007). Subsequently, mice with epidermal keratinocyte-specific gene alternations facilitated the study of the functional role of NF-κB pathways in the epidermal compartment. The generation of mice lacking IκBα in the epidermis further elucidated the crucial role of epidermal NF-κB. Simultaneous deletion of IκBα in epidermal keratinocytes and T cells resulted in inflammatory skin disease, indicating a T cell complemented role of IκBα in the epidermis. Furthermore, the deletion RelA in the epidermis fully prevented the development of skin lesions in IκBα^{E-KO} mice showing that increased NF-κB signaling in keratinocytes drives skin inflammation in this model (Rebholz et al. 2007).

Transgenic mice overexpressing degradation resistant super-repressor of IκBα (K5-IκBαSR and K14-IκBαM) showed that NF-κB inhibition also affects epidermal homeostasis. Both K5-IκBαSR and K14-IκBαM mice, which have strongly reduced NF-κB activity in the keratinocytes, showed increased keratinocyte proliferation and epidermal hyperplasia. K14-IκBαM mice died within 5–7 days of birth, whereas K5-IκBαSR mice survived until adulthood and developed squamous cell carcinoma in the FVB genetic background (Seitz et al. 1998; van Hogerlinden et al. 1999). K5-IκBαSR mice exhibited increased expression of TNF and death of keratinocytes, and did not develop skin lesions when bred with TNFR1-deficient mice (van Hogerlinden et al. 1999; Lind et al. 2004).

The generation and analysis of mice lacking IKK2 in keratinocytes (IKK2^{E-KO}) revealed that NF-κB inhibition in the epidermis is a potent trigger of skin inflammation. IKK2^{E-KO} mice were indistinguishable from their littermate controls at birth and started to develop skin lesions from postnatal day 4 (P4) onward. The skin lesions were characterized by epidermal hyperplasia, altered keratinocyte differentiation, upregulation of pro-inflammatory factors, and recruitment of immune cells. The skin lesions progressively developed to a very severe inflammatory skin condition resulting in the death of the mice before postnatal day 10 (Pasparakis et al. 2002). The inflammatory skin disease developing in IKK2^{E-KO} mice macroscopically and histologically resembles human psoriatic skin lesions. Moreover, the upregulation of signature genes, such as IL-1β, IL-6, IL-20, IL-19, and IL-24 in the epidermis of IKK2^{E-KO} mice resembled the expression pattern observed in human psoriatic epidermis (Kumari et al. 2013). Skin inflammation in IKK2^{E-KO} mice developed independently from T cells and was mainly driven by skin macrophages (Pasparakis et al. 2002; Stratis et al. 2006).

Anti-TNF neutralizing antibodies are highly effective in the treatment of psoriasis (Leonardi et al. 2003; Lowes et al. 2007). Similarly, skin inflammation in IKK2^{E-KO} mice was fully dependent on TNFR1 signaling, as full body or epidermal keratinocyte-specific ablation of TNFR1 prevented the development of skin lesions (Pasparakis et al. 2002; Kumari et al. 2013). These results showed that TNFR1 signaling in keratinocytes drives skin inflammation in IKK2^{E-KO} mice.

Further experiments identified the induction of IL-24 expression as a key mechanism by which epidermal TNFR1 signaling triggers inflammation in IKK2^{E-KO} mice. IL-24 was found to be overexpressed in the epidermis of IKK2^{E-KO} mice early on, followed by the upregulation of a number of inflammatory cytokines and chemokines. TNF-induced increased production of IL-24 in IKK2-deficient primary mouse keratinocytes or in primary human keratinocytes treated with an IKK2 inhibitor. Moreover, IL-24 expression in the epidermis was normalized in IKK2^{E-KO}/TNFR1^{E-KO} mice and also in psoriasis patients treated with anti-TNF antibodies. Finally, genetic deficiency of IL22R1, the main receptor of IL-24 strongly delayed and ameliorated skin lesion development in IKK2^{E-KO} mice providing functional evidence for the important role of IL-24 in driving skin inflammation (Kumari et al. 2013). Small numbers of dying keratinocytes are detected in the epidermis of IKK2^{E-KO} mice, but it remains unclear whether cell death plays a role in the development of skin inflammation in these animals.

Deficiency in the X chromosome-linked gene encoding NEMO/IKKγ, the regulatory component of the IKK complex, also caused skin lesion development in both humans and mice. NEMO deficiency in humans causes incontinentia pigmenti, a genetic disorder characterized by male embryonic lethality and the development of skin lesions and other complications in heterozygous females (Smahi et al. 2000). NEMO-deficient mice recapitulated the main features of human incontinentia pigmenti (Makris et al. 2000; Schmidt-Supprian et al. 2000). Mice with epidermis-specific knockout of NEMO also developed skin lesions, which were different to those of IKK2^{E-KO} mice (Nenci et al. 2006). Skin alterations are visible in NEMO^{E-KO} mice from P2 onward, mainly characterized by skin hypopigmentation and the subsequent development of severe skin inflammation leading to the death of the mice by P6–P7. The epidermis of NEMO^{E-KO} mice contained a large number of TUNEL-positive keratinocytes suggesting that death of NEMO-deficient keratinocytes is implicated in driving the skin pathology. NEMO-deficient keratinocytes were highly sensitive to TNF-induced death in vitro, and TNFR1 deficiency strongly delayed skin lesion development in NEMO^{E-KO} mice in vivo. However, NEMO^{EKO}/*Tnfr1*$^{-/-}$ mice developed inflammatory skin lesions later in life suggesting that the skin pathology depends on TNFR1 early on but is driven by TNFR1-independent signals at later stages (Nenci et al. 2006). The functional contribution of specific cell death pathways in the skin lesions developing in mice and humans with NEMO deficiency remains to be investigated.

2.2 *FADD/Caspase-8 in Epithelial Cells Prevent Necroptosis and Inflammation*

Keratinocyte-specific deletion of FADD (FADD^{E-KO}) or caspase-8 (CASP8^{E-KO}) resulted in severe skin inflammatory disease in mice, revealing an important role of FADD and caspase-8 in skin homeostasis and inflammation. Lee et al. reported that

CASP8^{E-KO} mice developed severe inflammatory skin lesions resulting in death of the mice by P10 (Lee et al. 2009). The authors of this study suggested that p38-dependent expression of IL-1α drives skin inflammation in CASP8^{E-KO} mice (Lee et al. 2009). Subsequently, Kovalenko et al. (2009) also reported that CASP8^{E-KO} mice develop skin inflammation. However, in contrast to Lee et al. Kovalenko et al. showed that the genetic ablation of IL-1α and IL-1β as well as pharmacological inhibition of p38 did not affect skin inflammation development in CASP8^{E-KO} mice and proposed an alternative model suggesting that an enhanced response to endogenous activators of retinoic acid inducible gene—Interferon regulatory factor 3 (RIG-I-IRF3) signaling in the epidermis, presumably generated in association with keratinocyte differentiation, contributes to the skin inflammatory process triggered by caspase-8 deficiency (Kovalenko et al. 2009; Rajput et al. 2011). However, Weinlich et al. showed later that the deficiency of mitochondrial antiviral-signaling protein (MAVS) did not prevent skin inflammation caused by inducible ablation of caspase-8 in the skin (Weinlich et al. 2013), suggesting that the RIG-I signaling axis does not play an important role in triggering skin inflammation in this model.

FADD^{E-KO} mice developed inflammatory skin lesions similarly to CASP8^{E-KO} mice (Bonnet et al. 2011). Genetic ablation of IL-1R or p38α did not prevent skin lesion development in FADD^{E-KO} mice, in agreement with the findings of Kovalenko et al. in CASP8^{E-KO} mice. Dying keratinocytes with necrotic appearance were detected in the epidermis of FADD^{E-KO} mice early on suggesting that keratinocyte necroptosis might contribute to the development of skin inflammation in these animals. Indeed, RIPK3 deficiency fully prevented the development of skin lesions in FADD^{E-KO} mice, providing genetic evidence that necroptosis of FADD-deficient keratinocytes drives skin inflammation. Subsequently, Weinlich et al. (2013) also showed that RIPK3 deficiency prevented skin lesion development in mice with inducible caspase-8 knockout in the epidermis, providing evidence that similarly to FADD^{E-KO} mice, RIPK3-dependent necroptosis triggers skin inflammation induced by caspase-8 deficiency. In both CASP8^{E-KO} and FADD^{E-KO} mice, TNF-TNFR1 signaling was identified as an important upstream pathway driving inflammation during the early stages. However, the deficiency of TNF or TNFR1 could not prevent skin lesion development at later stages, indicating involvement of TNFR1-independent signaling in driving RIPK3-dependent keratinocyte necroptosis and inflammation in CASP8^{E-KO} and FADD^{E-KO} mice. The lack of Fas, another death receptor, could also not prevent skin inflammation in FADD^{E-KO} mice (Kovalenko et al. 2009; Bonnet et al. 2011). It remains unclear whether death receptors have redundant roles in driving keratinocyte necroptosis and skin inflammation in CASP8^{E-KO} and FADD^{E-KO} mice, or other death receptor-independent pathways are also involved. TRIF-dependent TLR3/TLR4 signaling as well as IFNAR1 signaling has been shown to drive necroptosis, suggesting that these receptors could also be implicated in the development of skin lesions in CASP8^{E-KO} and FADD^{E-KO} mice.

2.3 CFLIP Contributes to Epidermal Homeostasis

Cellular FLICE-like inhibitory protein (cFLIP) is a component of death receptor-induced apoptotic complex that forms a heterodimeric complex with pro-caspase-8 and inhibits its effector function. Epidermis-specific inducible deletion of cFLIP resulted in inflammatory skin disease is characterized by epidermal hyperplasia, infiltration of immune cells, and death of keratinocytes (Panayotova-Dimitrova et al. 2013). RIPK3 deficiency did not prevent skin lesion development induced by cFLIP ablation in keratinocytes demonstrating that necroptosis is not required for the pathogenesis of the skin lesions (Weinlich et al. 2013). Since cFLIP inhibits both apoptosis and necroptosis, the development of skin lesions in mice with epidermis-specific cFLIP ablation is most likely driven by keratinocyte apoptosis. Dying keratinocytes in mice with epidermis-specific cFLIP ablation were positive for activated caspase-3 further supporting that keratinocyte apoptosis induces skin inflammation in this model. Indeed, the embryonic lethality of full body cFLIP knockout mice was rescued by double deficiency of FADD and RIPK3, indicating that cFLIP deficiency sensitizes cells to both apoptosis and necroptosis and that inhibition of both forms of cell death is required to prevent the pathologies induced by cFLIP deficiency. cFLIP-deficient keratinocytes were sensitive to TNF-induced apoptosis, while in vivo TNF neutralization using TNFR2-Fc partially ameliorated the phenotype of epidermal cFLIP-deficient mice (Weinlich et al. 2013), suggesting that TNF is an important mediator of the pathology.

2.4 RIP Kinase 1 Protects Necroptosis of Keratinocytes

Receptor interacting kinase 1 (RIPK1) regulates both cell survival and programmed cell death by apoptosis and necroptosis. Mice with epidermis-specific deletion of RIPK1 ($RIPK1^{E-KO}$) developed skin lesion starting from P8 that progressed over time resulting in severe inflammatory skin disease by P21–P28. The skin lesions of $RIPK1^{E-KO}$ are characterized by epidermal hyper-proliferation, altered differentiation, and upregulation of pro-inflammatory factors in the skin. Immunostaining for activated caspase-3 showed the presence of small numbers of apoptotic keratinocytes in the epidermis of $RIPK1^{E-KO}$ mice; however, some of the dying keratinocytes did not stain for activated caspase-3 indicating they did not die by apoptosis. These findings suggested that keratinocyte death by apoptosis or necroptosis could be implicated in the pathology. Indeed, genetic ablation of RIPK3 or MLKL fully prevented skin lesion development in $RIPK1^{E-KO}$ mice demonstrating that necroptosis of RIPK1-deficient keratinocytes drives skin inflammation in this model (Dannappel et al. 2014). Moreover, mice with RIPK1 deficiency in all cells die perinatally exhibiting multiple abnormalities including epidermal hyperplasia and skin inflammation. Interestingly, RIPK3 or MLKL deficiency could prevent skin hyperplasia in newborn *Ripk1*$^{-/-}$ mice, further supporting a role of RIPK1 as an inhibitor of necroptosis that regulates skin homeostasis (Dillon et al. 2014; Rickard et al. 2014b). These studies identified a unique physiological role of RIPK1 as negative regulator of RIPK3 that prevents necroptosis of keratinocytes and maintains skin homeostasis. Unlike RIPK1^{E-KO} mice, mutant mice expressing an inactive kinase RIPK1 allele did not develop skin lesions showing that the protective role of RIPK1 in keratinocyte death and skin inflammation is due to its kinase-independent scaffolding function (Dannappel et al. 2014; Polykratis et al. 2014). TNFR1 deficiency delayed and partially ameliorated the skin lesions in RIPK1^{E-KO} mice, suggesting that TNF-independent pathways are also involved in inducing necroptosis of RIPK1-deficient keratinocytes. Therefore, in addition to the FADD^{E-KO} mice, RIPK1^{E-KO} mice provided genetic evidence that keratinocyte necroptosis potently induces skin inflammation.

2.5 SHARPIN Inhibits Keratinocytes Apoptosis

Sharpin is a component of LUBAC and is involved in the linear ubiquitination of NEMO and RIPK1 downstream of TNFR1 (Gerlach et al. 2011; Ikeda et al. 2011; Tokunaga et al. 2011). A spontaneous mutation in the *Sharpin* gene caused the chronic proliferative dermatitis (cpdm) phenotype in mice, which is primarily characterized by the development of severe inflammatory skin lesions but also inflammation in other tissues. Dermatitis in *Sharpin*$^{cpdm/cpdm}$ mice is characterized by epidermal hyper-proliferation, immune cell infiltration in the dermis, and massive death of keratinocytes in the epidermis (Gijbels et al. 1996; Seymour et al. 2007). Most dying keratinocytes in *Sharpin*$^{cpdm/cpdm}$ mice are positive for activated caspase-3 suggesting that apoptosis is the primary mode of keratinocyte death in these animals. Indeed, *Sharpin*$^{cpdm/cpdm}$ mice lacking RIPK3 or MLKL showed a delayed onset of the dermatitis and partial amelioration of the multi-organ phenotype, indicating a relatively small contribution of necroptosis to the phenotype (Kumari et al. 2014; Rickard et al. 2014a). On the other hand, the deletion of FADD in the epidermis combined with systemic RIPK3 deficiency completely prevented keratinocyte death and skin inflammation in *Sharpin*$^{cpdm/cpdm}$ mice, indicating that FADD-mediated death of keratinocytes is the main trigger for the dermatitis in *Sharpin*$^{cpdm/cpdm}$ mice (Kumari et al. 2014). Consistent with these findings, combined RIPK3 deficiency with heterozygous knockout of caspase-8 also fully prevented skin inflammation in *Sharpin*$^{cpdm/cpdm}$ mice, further supporting that FADD/caspase-8-mediated apoptosis of Sharpin-deficient keratinocytes plays a key role in triggering inflammation in these mice (Rickard et al. 2014a). Furthermore, the death of keratinocytes as well as skin inflammation in *Sharpin*$^{cpdm/cpdm}$ mice was completely prevented by systemic deletion of TNF or TNFR1 in the full body (Gerlach et al. 2011; Kumari et al. 2014; Rickard et al. 2014a) but also by keratinocyte-specific TNFR1 knockout (Kumari et al. 2014). Epidermis-specific

TNFR1 deficiency prevented inflammation in the skin but did not affect inflammation in other organs of *Sharpin*$^{cpdm/cpdm}$ mice, indicating that cell intrinsic TNFR1 signaling drives inflammation in the different tissues of these animals (Kumari et al. 2014).

3 Conclusion and Perspectives

Studies in genetically modified mouse models provide ample evidence that keratinocytes are key players in maintaining skin homeostasis. Signaling pathways regulating keratinocyte responses to stress and danger signals play an important role in the regulation of physiological and pathological processes in the skin. Considering that due to their strategically essential position in the outermost barrier of the body keratinocytes are challenged every day by a plethora of mechanical, biological, and chemical insults, their capacity to rapidly respond to these signals and communicate this information to immune cells is probably an evolutionary adaptation allowing the rapid detection and response to danger coming from the environment. While this function has primarily beneficial effects in host defense and wound healing, deregulation of these responses results in uncontrolled inflammation triggering the pathogenesis of chronic inflammatory skin diseases.

It is interesting to note that the compromised skin homeostasis in the different mouse models shares similar characteristics manifesting with epidermal hyperplasia, accumulation of immune cells, upregulation of pro-inflammatory markers, and often death of keratinocytes (apoptosis and/or necroptosis). The coexistence of inflammation and keratinocyte death in these skin lesions suggests that these processes are linked. Experimental evidence in some of the mouse models suggests that the death of keratinocytes provides the trigger initiating skin inflammation; however in other cases, it is unclear whether cell death is the cause or the consequence of the inflammatory response. Skin inflammation is coordinated by the expression of multiple pro-inflammatory factors, which trigger activation of keratinocytes as well as activation and recruitment of immune cells. A key question that remains to be experimentally addressed is how the death of keratinocytes could trigger skin inflammation. Several plausible mechanisms can be proposed for the inter-dependent co-existence of inflammation and keratinocyte death in the skin.

Cell death likely provides a protective mechanism in keratinocytes, which eliminates the stimulus-affected cells. The epidermis is daily exposed to potentially hazardous insults from the environment, and therefore, the elimination of the damaged keratinocytes could be important to maintain homeostasis and prevent the propagation of intracellular pathogens. Moreover, death of keratinocytes exposed to genotoxic stress could be important to prevent skin cancer. Myeloid cells accumulate in the epidermis to clear the dead cells triggering local inflammation. Under healthy homeostatic conditions, this local inflammatory response is terminated when the threat has been eliminated allowing the restoration of normal skin homeostasis. However, when keratinocyte responses to stress are deregulated resulting in exaggerated cell death or/and cytokine secretion, these normally local and transient inflammatory responses could be prolonged and amplified resulting in chronic inflammation.

The important role of TNF not only in multiple inflammatory skin conditions in mouse models but also in human psoriasis identifies TNF/TNFR signaling as a key pathogenic factor. TNF is a potent inducer of inflammation via the activation of NF-κB-dependent expression of inflammatory genes, but can also potently trigger cell death under conditions where NF-κB signaling is compromised. It is therefore tempting to speculate that TNF-induced keratinocyte death might have evolved as a mechanism allowing the activation of an immune response under conditions of NF-κB inhibition. Considering that many bacteria and viruses have developed strategies to inhibit NF-κB in order to evade the immune system, cell death could provide a mechanism facilitating host defense to those pathogens. Also in this case, deregulation of keratinocyte responses to TNF could provide a mechanism triggering uncontrolled inflammation resulting in chronic skin diseases.

Another key question that remains to be addressed is the potential differential role of different modes of keratinocyte death in skin inflammation. Apoptosis is generally considered as non-immunogenic, whereas necroptosis is thought to constitute a more pro-inflammatory cell death modality by various means. Necroptotic keratinocytes can potentially release damage-associated molecular pattern molecules (DAMPs), e.g., high mobility group protein B1 (HMGB)-1, S100A8, S100A9, ATP, IL-33, and uric acid (Pasparakis and Vandenabeele 2015). These DAMPs are believed to alarm and activate the surrounding keratinocytes to circumvent the adverse situation and promote activation and recruitment of immune cells in the dermis (Fig. 1). However, the role of DAMPs produced by necroptotic cells in inflammation has been proposed on the basis of correlative evidence, as there is currently no experimental evidence demonstrating that the release of specific DAMPs by necroptotic cells is functionally involved in triggering inflammation. Providing experimental evidence demonstrating that the release of DAMPs by necroptotic cells is functionally important for inflammation will be essential in order to support this model. In contrast to necroptosis, apoptosis is mainly known for its good side in maintaining tissue homeostasis. Although the presence of apoptotic keratinocytes is reported in most of the mouse models of inflammatory skin diseases, so far keratinocyte apoptosis is identified as a trigger of inflammation only in *Sharpin*$^{cpdm/cpdm}$ and probably in cFLIP^{E-KO} mice. The question that arises from these studies is how could apoptosis drive skin inflammation? One possibility is that secondary necrosis of apoptotic cells perhaps caused by inefficient clearance of these cells could provide a signal for inflammation by allowing the release of DAMPS. In addition, caspase-8 has been implicated in the activation of IL-1β release either by directly processing pro-IL-1β or via the activation of caspase-1 (Maelfait et al. 2008; Bossaller et al. 2012; Vince et al. 2012; Antonopoulos et al. 2013; Gurung et al. 2014; Philip et al. 2014; Weng et al. 2014). This suggests that at least under certain conditions when the cell may be primed, activation of caspase-8 may result in immunogenic apoptotic cell death.

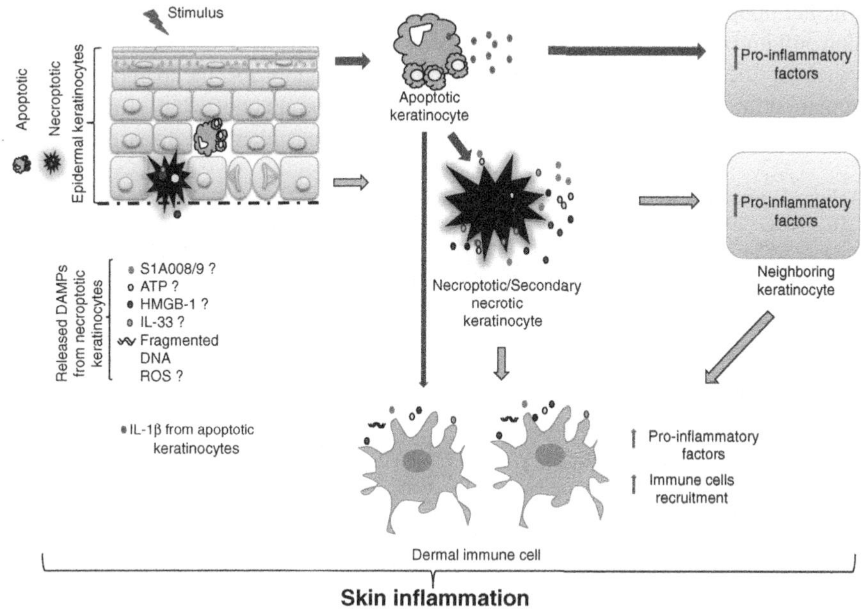

Fig. 1 Graphical representation of the potential mechanisms by which keratinocytes dying by apoptosis or necroptosis could trigger skin inflammation. *ROS* reactive oxygen species, *ATP* adenosine triphosphate, *IL* interleukin, *HMGB-1* high mobility group protein B-1

Collectively, data from a number of mouse models provided experimental evidence that keratinocyte death provides a potent trigger of skin inflammation. Is this relevant for human inflammatory skin diseases? Death of keratinocytes is observed in several human inflammatory skin diseases, e.g., oral lichen planus, pemphigus vulgaris, Hailey–Hailey disease, Grover's disease, Lupus, autoimmune disorders, TEN/Lyell syndrome, drug-induced toxicity, chronic wounds and Psoriasis has been sporadically reported (Wrone-Smith et al. 1995; Gniadecki et al. 1998; Kawashima et al. 2004). However, the functional role of keratinocyte death in the pathogenesis of these skin diseases is unknown. Moreover, most of the studies have used TUNEL as a marker to detect death of keratinocytes, which has been used for many years as a specific marker of apoptosis. However, TUNEL can also stain necrotic cells; therefore, a more detailed analysis of the type of cell death using specific markers of apoptosis or necroptosis would be required in order to define the type of keratinocyte death in different human skin diseases. Activated caspase-3 is a very good marker for the detection of apoptotic cells, but until recently there were no specific markers allowing the detection of necroptotic cells in tissue sections. The recent development of antibodies specifically recognizing phosphorylated human MLKL promises to provide a tool allowing the specific detection of necroptotic cells (Wang et al. 2014). The comparison of the type of keratinocyte death in different inflammatory skin diseases may provide important information on whether apoptosis and necroptosis have differential functions in the pathogenesis of skin inflammation. These studies will be important in order to evaluate the relevance of the knowledge obtained from mouse models for the pathogenesis of human skin diseases and to validate the potential therapeutic value of approaches targeting different cell death pathways in the treatment for these diseases.

Acknowledgments Research in the authors' laboratory is funded by the ERC (2012-ADG_20120314), the DFG (SFB670, SFB829, SPP1656), the European Commission [Grants 223404 (Masterswitch) and 223151 (InflaCare)], the Deutsche Krebshilfe, the Else Kröner-Fresenius-Stiftung and the Helmholtz Alliance (PCCC).

References

Antonopoulos C, El Sanadi C, Kaiser WJ et al (2013) Proapoptotic chemotherapeutic drugs induce noncanonical processing and release of IL-1beta via caspase-8 in dendritic cells. J Immunol 191(9):4789–4803

Beg AA, Sha WC, Bronson RT et al (1995) Constitutive NF-kappa B activation, enhanced granulopoiesis, and neonatal lethality in I kappa B alpha-deficient mice. Genes Dev 9 (22):2736–2746

Bonnet MC, Preukschat D, Welz PS et al (2011) The adaptor protein FADD protects epidermal keratinocytes from necroptosis in vivo and prevents skin inflammation. Immunity 35(4):572–582

Bossaller L, Chiang PI, Schmidt-Lauber C et al (2012) Cutting edge: FAS (CD95) mediates noncanonical IL-1beta and IL-18 maturation via caspase-8 in an RIP3-independent manner. J Immunol 189(12):5508–5512

Chen ZJ, Bhoj V, Seth RB (2006) Ubiquitin, TAK1 and IKK: is there a connection? Cell Death Differ 13(5):687–692

Cho YS, Challa S, Moquin D et al (2009) Phosphorylation-driven assembly of the RIP1-RIP3 complex regulates programmed necrosis and virus-induced inflammation. Cell 137(6):1112–1123

Dannappel M, Vlantis K, Kumari S et al (2014) RIPK1 maintains epithelial homeostasis by inhibiting apoptosis and necroptosis. Nature 513(7516):90–94

Di Meglio P, Perera GK, Nestle FO (2011) The multitasking organ: recent insights into skin immune function. Immunity 35(6):857–869

Dillon CP, Weinlich R, Rodriguez DA et al (2014) RIPK1 blocks early postnatal lethality mediated by caspase-8 and RIPK3. Cell 157(5):1189–1202

Gerlach B, Cordier SM, Schmukle AC et al (2011) Linear ubiquitination prevents inflammation and regulates immune signalling. Nature 471(7340):591–596

Gijbels MJ, Zurcher C, Kraal G et al (1996) Pathogenesis of skin lesions in mice with chronic proliferative dermatitis (cpdm/cpdm). Am J Pathol 148(3):941–950

Gniadecki R, Jemec GB, Thomsen BM et al (1998) Relationship between keratinocyte adhesion and death: anoikis in acantholytic diseases. Arch Dermatol Res 290(10):528–532

Gurung P, Anand PK, Malireddi RK et al (2014) FADD and caspase-8 mediate priming and activation of the canonical and noncanonical Nlrp3 inflammasomes. J Immunol 192(4):1835–1846

Haas AL (2009) Linear polyubiquitylation: the missing link in NF-kappaB signalling. Nat Cell Biol 11(2):116–118

Haas TL, Emmerich CH, Gerlach B et al (2009) Recruitment of the linear ubiquitin chain assembly complex stabilizes the TNF-R1 signaling complex and is required for TNF-mediated gene induction. Mol Cell 36(5):831–844

Hayden MS, Ghosh S (2008) Shared principles in NF-kappaB signaling. Cell 132(3):344–362

He S, Wang L, Miao L et al (2009) Receptor interacting protein kinase-3 determines cellular necrotic response to TNF-alpha. Cell 137(6):1100–1111

He S, Liang Y, Shao F et al (2011) Toll-like receptors activate programmed necrosis in macrophages through a receptor-interacting kinase-3-mediated pathway. Proc Natl Acad Sci U S A 108(50):20054–20059

Ikeda F, Deribe YL, Skanland SS et al (2011) SHARPIN forms a linear ubiquitin ligase complex regulating NF-kappaB activity and apoptosis. Nature 471(7340):637–641

Jacobson MD, Weil M, Raff MC (1997) Programmed cell death in animal development. Cell 88 (3):347–354

Kaiser WJ, Sridharan H, Huang C et al (2013) Toll-like receptor 3-mediated necrosis via TRIF, RIP3, and MLKL. J Biol Chem 288(43):31268–31279

Kawashima K, Doi H, Ito Y et al (2004) Evaluation of cell death and proliferation in psoriatic epidermis. J Dermatol Sci 35(3):207–214

Klement JF, Rice NR, Car BD et al (1996) IkappaBalpha deficiency results in a sustained NF-kappaB response and severe widespread dermatitis in mice. Mol Cell Biol 16(5):2341–2349

Kovalenko A, Kim JC, Kang TB et al (2009) Caspase-8 deficiency in epidermal keratinocytes triggers an inflammatory skin disease. J Exp Med 206(10):2161–2177

Kumari S, Bonnet MC, Ulvmar MH et al (2013) Tumor necrosis factor receptor signaling in keratinocytes triggers interleukin-24-dependent psoriasis-like skin inflammation in mice. Immunity 39(5):899–911

Kumari S, Redouane Y, Lopez-Mosqueda J et al (2014) Sharpin prevents skin inflammation by inhibiting TNFR1-induced keratinocyte apoptosis. eLife 3:e03422

Lee P, Lee DJ, Chan C et al (2009) Dynamic expression of epidermal caspase 8 simulates a wound healing response. Nature 458(7237):519–523

Leonardi CL, Powers JL, Matheson RT et al (2003) Etanercept as monotherapy in patients with psoriasis. N Engl J Med 349(21):2014–2022

Lind MH, Rozell B, Wallin RP et al (2004) Tumor necrosis factor receptor 1-mediated signaling is required for skin cancer development induced by NF-kappaB inhibition. Proc Natl Acad Sci U S A 101(14):4972–4977

Lippens S, Kockx M, Knaapen M et al (2000) Epidermal differentiation does not involve the pro-apoptotic executioner caspases, but is associated with caspase-14 induction and processing. Cell Death Differ 7(12):1218–1224

Lippens S, Denecker G, Ovaere P et al (2005) Death penalty for keratinocytes: apoptosis versus cornification. Cell Death Differ 12(Suppl 2):1497–1508

Lowes MA, Bowcock AM, Krueger JG (2007) Pathogenesis and therapy of psoriasis. Nature 445 (7130):866–873

Maelfait J, Vercammen E, Janssens S et al (2008) Stimulation of Toll-like receptor 3 and 4 induces interleukin-1beta maturation by caspase-8. J Exp Med 205(9):1967–1973

Makris C, Godfrey VL, Krahn-Senftleben G et al (2000) Female mice heterozygous for IKK gamma/NEMO deficiencies develop a dermatopathy similar to the human X-linked disorder incontinentia pigmenti. Mol Cell 5(6):969–979

McComb S, Cessford E, Alturki NA et al (2014) Type-I interferon signaling through ISGF3 complex is required for sustained Rip3 activation and necroptosis in macrophages. Proc Natl Acad Sci U S A 111(31):E3206–E3213

Milligan CE, Schwartz LM (1997) Programmed cell death during animal development. Br Med Bull 53(3):570–590

Murphy JM, Czabotar PE, Hildebrand JM et al (2013) The pseudokinase MLKL mediates necroptosis via a molecular switch mechanism. Immunity 39(3):443–453

Nenci A, Huth M, Funteh A et al (2006) Skin lesion development in a mouse model of incontinentia pigmenti is triggered by NEMO deficiency in epidermal keratinocytes and requires TNF signaling. Hum Mol Genet 15(4):531–542

Nestle FO, Di Meglio P, Qin JZ et al (2009) Skin immune sentinels in health and disease. Nat Rev Immunol 9(10):679–691

Panayotova-Dimitrova D, Feoktistova M, Ploesser M et al (2013) cFLIP regulates skin homeostasis and protects against TNF-induced keratinocyte apoptosis. Cell Rep 5(2):397–408

Pasparakis M, Vandenabeele P (2015) Necroptosis and its role in inflammation. Nature 517 (7534):311–320

Pasparakis M, Courtois G, Hafner M et al (2002) TNF-mediated inflammatory skin disease in mice with epidermis-specific deletion of IKK2. Nature 417(6891):861–866

Pasparakis M, Haase I, Nestle FO (2014) Mechanisms regulating skin immunity and inflammation. Nat Rev Immunol 14(5):289–301

Philip NH, Dillon CP, Snyder AG et al (2014) Caspase-8 mediates caspase-1 processing and innate immune defense in response to bacterial blockade of NF-kappaB and MAPK signaling. Proc Natl Acad Sci U S A 111(20):7385–7390

Polykratis A, Hermance N, Zelic M et al (2014) Cutting edge: RIPK1 Kinase inactive mice are viable and protected from TNF-induced necroptosis in vivo. J Immunol 193(4):1539–1543

Rajput A, Kovalenko A, Bogdanov K et al (2011) RIG-I RNA helicase activation of IRF3 transcription factor is negatively regulated by caspase-8-mediated cleavage of the RIP1 protein. Immunity 34(3):340–351

Rebholz B, Haase I, Eckelt B et al (2007) Crosstalk between keratinocytes and adaptive immune cells in an IκBα protein-mediated inflammatory disease of the skin. Immunity 27(2):296–307

Rickard JA, Anderton H, Etemadi N et al (2014a) TNFR1-dependent cell death drives inflammation in Sharpin-deficient mice. eLife 3:e03464

Rickard JA, O'Donnell JA, Evans JM et al (2014b) RIPK1 regulates RIPK3-MLKL-driven systemic inflammation and emergency hematopoiesis. Cell 157(5):1175–1188

Schmidt-Supprian M, Bloch W, Courtois G et al (2000) NEMO/IKK gamma-deficient mice model incontinentia pigmenti. Mol Cell 5(6):981–992

Seitz CS, Lin Q, Deng H et al (1998) Alterations in NF-kappaB function in transgenic epithelial tissue demonstrate a growth inhibitory role for NF-kappaB. Proc Natl Acad Sci U S A 95 (5):2307–2312

Seymour RE, Hasham MG, Cox GA et al (2007) Spontaneous mutations in the mouse Sharpin gene result in multiorgan inflammation, immune system dysregulation and dermatitis. Genes Immun 8(5):416–421

Smahi A, Courtois G, Vabres P et al (2000) Genomic rearrangement in NEMO impairs NF-kappaB activation and is a cause of incontinentia pigmenti. The International Incontinentia Pigmenti (IP) Consortium. Nature 405(6785):466–472

Stratis A, Pasparakis M, Rupec RA et al (2006) Pathogenic role for skin macrophages in a mouse model of keratinocyte-induced psoriasis-like skin inflammation. J Clin Invest 116(8):2094–2104

Sun L, Wang H, Wang Z et al (2012) Mixed lineage kinase domain-like protein mediates necrosis signaling downstream of RIP3 kinase. Cell 148(1–2):213–227

Thapa RJ, Nogusa S, Chen P et al (2013) Interferon-induced RIP1/RIP3-mediated necrosis requires PKR and is licensed by FADD and caspases. Proc Natl Acad Sci U S A 110(33): E3109–E3118

Tokunaga F, Nakagawa T, Nakahara M et al (2011) SHARPIN is a component of the NF-kappaB-activating linear ubiquitin chain assembly complex. Nature 471(7340):633–636

Upton JW, Kaiser WJ, Mocarski ES (2012) DAI/ZBP1/DLM-1 complexes with RIP3 to mediate virus-induced programmed necrosis that is targeted by murine cytomegalovirus vIRA. Cell Host Microbe 11(3):290–297

van Hogerlinden M, Rozell BL, Ahrlund-Richter L et al (1999) Squamous cell carcinomas and increased apoptosis in skin with inhibited Rel/nuclear factor-kappaB signaling. Cancer Res 59 (14):3299–3303

Vanden Berghe T, Linkermann A, Jouan-Lanhouet S et al (2014) Regulated necrosis: the expanding network of non-apoptotic cell death pathways. Nat Rev Mol Cell Biol 15 (2):135–147

Vandenabeele P, Declercq W, Van Herreweghe F et al (2010) The role of the kinases RIP1 and RIP3 in TNF-induced necrosis. Sci Signal 3(115):re4

Vince JE, Wong WW, Gentle I et al (2012) Inhibitor of apoptosis proteins limit RIP3 kinase-dependent interleukin-1 activation. Immunity 36(2):215–227

Wang H, Sun L, Su L et al (2014) Mixed lineage kinase domain-like protein MLKL causes necrotic membrane disruption upon phosphorylation by RIP3. Mol Cell 54(1):133–146

Weinlich R, Oberst A, Dillon CP et al (2013) Protective roles for caspase-8 and cFLIP in adult homeostasis. Cell Rep 5(2):340–348

Weng D, Marty-Roix R, Ganesan S et al (2014) Caspase-8 and RIP kinases regulate bacteria-induced innate immune responses and cell death. Proc Natl Acad Sci U S A 111 (20):7391–7396

Wilson NS, Dixit V, Ashkenazi A (2009) Death receptor signal transducers: nodes of coordination in immune signaling networks. Nat Immunol 10(4):348–355

Wrone-Smith T, Johnson T, Nelson B et al (1995) Discordant expression of Bcl-x and Bcl-2 by keratinocytes in vitro and psoriatic keratinocytes in vivo. Am J Pathol 146(5):1079–1088

Zhang DW, Shao J, Lin J et al (2009) RIP3, an energy metabolism regulator that switches TNF-induced cell death from apoptosis to necrosis. Science 325(5938):332–336

Zhao J, Jitkaew S, Cai Z et al (2012) Mixed lineage kinase domain-like is a key receptor interacting protein 3 downstream component of TNF-induced necrosis. Proc Natl Acad Sci U S A 109(14):5322–5327

IAPs and Cell Death

John Silke and James Vince

Abstract IAPs were named as inhibitors of apoptosis, programmed cell death, but it has become apparent that they are regulators of other types of cell death too. Because they inhibit cell death in cancer cells there has been an intense interest in developing inhibitors of these proteins to induce or sensitise cancer cells to death. In this article, we will discuss the involvement of IAPs in the apoptosis, necroptosis and pyroptosis programmed cell death paradigms. All these types of cell death are intimately involved with causing or repressing inflammation and it should perhaps therefore come as no surprise that IAPs are also involved in regulating inflammation directly. To come full circle, the IAP antagonist drugs that were developed to sensitise cancer cells to apoptosis have led to some of these insights.

Contents

J. Silke (✉)
The Walter and Eliza Hall Institute of Medical Research, Melbourne, VIC 3052, Australia
e-mail: silke@wehi.edu.au; j.silke@latrobe.edu.auor

J. Vince
Department of Medical Biology, University of Melbourne, Melbourne, Australia
e-mail: vince@wehi.edu.au

Current Topics in Microbiology and Immunology (2017) 403:95–118
DOI 10.1007/82_2016_507

Published Online: 14 February 2017

1 Introduction

IAPs were so named because they were identified in a screen where they inhibited insect cell death caused by a baculovirus. Since then many IAPs have been identified and our understanding of their function has become much more nuanced and it is clear that they regulate signalling from a host of receptors that signal in the innate and adaptive immune response primarily through the intermediary of the multi-talented workhorse, ubiquitin. As the subject of this review is IAPs and cell death, we will endeavour to focus on the cell death regulation aspect and signalling where it intersects with this topic. As can be appreciated the boundaries are not always sharp and the field is by now very large so we have had to exercise our own idiosyncratic judgment on what to discuss and what not. IAPs all share the defining Baculoviral IAP repeat (BIR) and many also contain a RING finger domain that allows them to utilise ubiquitin. However, rather than taking the standard approach of describing the domains of IAPs (Fig. 1) which inevitably leads to discussions of function (Silke and Meier 2013), we have taken an historical approach, with particular reference to studies on what we consider to be the three main cell death regulating mammalian IAPs. It will not simply be a timeline however as we will try and finish one discussion thread before tackling the next one in order to avoid excessive switching between different themes. We conclude with some outstanding questions.

2 IAPs and Apoptosis

2.1 The Very Beginning—How IAPs Were Discovered

IAPs were first discovered in baculoviruses, viruses that infect invertebrate cells and which were early on investigated for their potential as natural insecticides (Bird 1964; Inceoglu et al. 2006), but which are now mostly used as recombinant protein expression vectors. In this age of the translation mantra it is well worth reminding funding agencies that this discovery, and no doubt these other uses, was driven and enabled overall by curiosity rather than its translation potential. These first baculoviral IAPs, OpIAP and CpIAP, were identified in the genomes of baculoviruses CpGV and OpNPV because of their ability to rescue deficiency of a baculoviral caspase inhibitor called p35 (Crook et al. 1993; Birnbaum et al. 1994). For a recent history of the baculoviral IAPs the interested reader is directed to an article by Clem (2015).

Mammalian cellular IAPs (cIAPs) were first identified binding to TNFR2 indirectly through the intermediary of TRAF1 and TRAF2 (Rothe et al. 1995). It was immediately recognised that these proteins bore homology to the baculoviral IAPs, in their multiple BIR and their RING fingers (Fig. 1a). However, since it was not

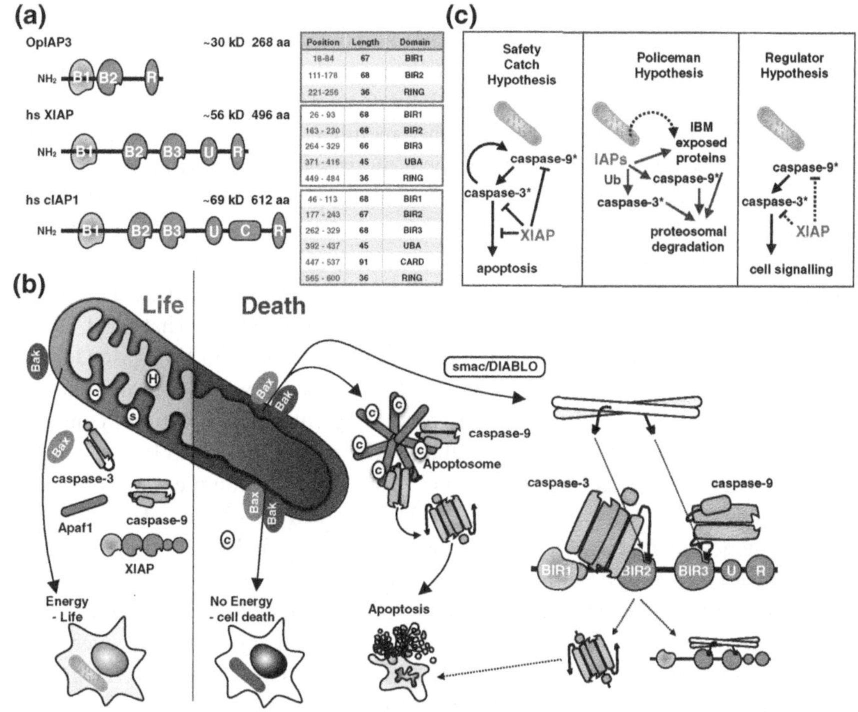

Fig. 1 **a** A schematic representation of the domain structure of a baculoviral IAP and human XIAP and cIAP1. **b** Mechanisms of death and interplay of XIAP. c; cytochrome c. s; Smac/DIABLO and H; HtrA2 are two examples of IAP antagonist molccules present in mitochondria that are released following disruption of the outer mitochondrial membrane. The key points are that, usually, activation of Bax and or Bak should be sufficient to induce cell death and that activation of caspase-9 and caspase-3 performs the function of promoting *apoptotic* cell death. Note that the apoptosome most likely contains cytochrome c and caspase-9 bound to all 7 spokes of the apoptosome wheel but they have not all been shown to simplify the figure. **c** Three major hypotheses discussed in the main text proposed to rationalise the caspase inhibitory activity of XIAP. In the safety catch hypothesis XIAP is able to inhibit (and probably ubiquitylate and promote proteosomal degradation) inadvertently activated caspases preventing accidental death. The policeman hypothesis is an extended variation of the safety catch hypothesis and includes proteins that have an exposed IAP Binding Motif (IBM) potentially because they have been accidentally released from mitochondria or another cellular compartment normally inaccessible to any of the IAPs. The regulator hypothesis proposes that there are some situations, presumably independent of mitochondrial permeabilisation, where caspase-9 or caspase-3 are activated but their activity is channelled or modified to specific substrates such that their activity does not induce apoptosis

known how these baculoviral proteins inhibited apoptosis, nor what the BIR and RING domains might do, this did not lead to much insight into their function at the TNFR2 receptor. Subsequently it was shown that the IAPs bind to TRAFs via their BIR1 domains (Samuel et al. 2006; Varfolomeev et al. 2006). The motif within TRAF2 required for cIAP interaction was then identified (Vince et al. 2009) and the crystal structures of the interacting regions were solved (Zheng et al. 2010; Mace et al. 2010). Surprisingly, the structures and biophysical analyses revealed that one IAP molecule binds to a TRAF2 trimer (Zheng et al. 2010; Mace et al. 2010), and that the BIR1 of cIAP binds preferentially to a $TRAF2^2.TRAF1^1$ heterotrimer (Zheng et al. 2010). TNFR2, like 20 other members of the TNF Receptor superfamily (TNFRSF) has a TRAF2 recruitment motif and it is not surprising therefore that cIAPs regulate signalling from this class of receptors (Ye et al. 1999; Park et al. 1999; Silke and Brink 2010; Varfolomeev et al. 2012). Even if, like TNFR1, the receptors do not have TRAF2 binding sites, but are able to recruit the TRAF2 interacting protein, TRADD (Park et al. 2000), then they will also be able to recruit TRAFs and cIAPs. Once recruited however the question remains what do cIAPs do? The RING domain was shown in 1999 to function as a ubiquitin E3 ligase (Joazeiro et al. 1999) and shortly after the RINGs of IAPs were shown to contribute to their proteasomal degradation presumably by promoting auto-ubiquitylation (Yang et al. 2000). In TNFR1 signalling, it appears that cIAPs are the critical E3 ligases required to ubiquitylate RIPK1 and probably other proteins in the complex. This has been shown using *cIap1*$^{-/-}$.*cIap2*$^{-/-}$ cells, cells in which cIAPs are depleted using Smac-mimetics or shRNA or combinations of knock-out and knock-down and in all cases RIPK1 ubiquitylation was almost completely lost (Haas et al. 2009; Varfolomeev et al. 2008; Mahoney et al. 2008). Loss of RIPK1 ubiquitylation coincides with impaired activation of MAPK and NF-κB signalling however loss of RIPK1 does not always result in loss of TNF-induced MAPK and NF-κB signalling (Wong et al. 2010; Dannappel et al. 2014; Takahashi et al. 2014; Haas et al. 2009; Blackwell et al. 2013). This therefore suggests that other components within the signalling complex can serve as substrates for cIAP mediated ubiquitylation (Blackwell et al. 2013). Furthermore while TRAF2 appears to be the major recruitment partner for cIAPs, *Traf2*$^{-/-}$ cells are rarely completely defective in their ability to activate NF-κB or JNK (Lee et al. 1997; Yeh et al. 1997; Vince et al. 2009; Zhang et al. 2016; Etemadi et al. 2015). Again this suggests that cIAPs can be recruited into the TNFR1 signalling complex by other interacting partners in addition to TRAF2. TRAF5 might be one of those factors because *Traf2*$^{-/-}$.*Traf5*$^{-/-}$ double knock-out cells are more defective than *Traf2*$^{-/-}$ cells in their ability to activate NF-κB (Tada et al. 2001; Vince et al. 2009), however TRAF5 does not contain an IAP interacting motif (Gentle and Silke 2011) and TRAF5 may not be as important in many cells in TNF-induced signalling as was first assumed (Zhang et al. 2016). How this ubiquitin-dependent signalling affects cell survival in response to TNFR1 will be dealt with in a subsequent section.

2.2 XIAP Inhibiting Apoptosis

XIAP received the lion's share of attention in the early days of IAP research. Several groups had shown that XIAP, (called variously hILP, MIHA), like its baculoviral homologs, had the ability to inhibit cell death and particularly caspase-mediated apoptosis (Uren et al. 1996; Duckett et al. 1996; Liston et al. 1996). However a key breakthrough was the finding that XIAP was directly able to inhibit cytochrome c induced activation of caspase-3 and bind and inhibit caspase-3 itself (Deveraux et al. 1997; Fig. 1b). Consistent with data showing that cytochrome c activates the upstream initiator caspase-9 (Li et al. 1997), it was shown soon after that XIAP had the ability to bind to pro-caspase-9 and inhibit activated caspase-9 in vitro (Deveraux et al. 1998; Fig. 1b). An XIAP BIR2 with flanking regions was the minimum domain required to inhibit caspase-3 and caspase-7 (Takahashi et al. 1998) while a different BIR, BIR3 was required to inhibit caspase-9 (Deveraux et al. 1999). Surprisingly, particularly given the degree of amino acid identity of the BIRs and the homology of the caspases themselves, the mechanism of action of these two different domains turned out to be distinct. While the BIR2 domain is required to bind processed caspase-3 the actual inhibitory domain lies in the N-terminal linker that joins BIR1 and BIR2 (Sun et al. 1999). Caspases are so called because they are cysteine proteases that cleave after Aspartate. Furthermore downstream caspases such as caspase-3 have a preferred DXXD cleavage site. Therefore when Asp148 within the BIR1 BIR2 linker and part of the tetrapeptide DISD was shown to be critical for caspase inhibition, it was assumed that the mechanism of action might be similar to the viral caspase inhibitors CrmA and p35 that function as pseudo- or suicide substrate inhibitors (Zhou et al. 1998; Fisher et al. 1999). However this assumption turned out to be false; in fact the linker interacts in a C–N orientation with the active site of caspase-3 and this renders it uncleavable (Silke et al. 2001; Huang et al. 2001; Riedl et al. 2001; Chai et al. 2001; Silke and Meier 2013). The mechanism of action of the BIR3 of XIAP against caspase-9 is quite different. Because caspase-9 needs to oligomerise in order to become, and be, active (Boatright et al. 2003), rather than be processed, mechanistically it makes sense that XIAP inhibits caspase-9 in a different manner to caspase-3. Nevertheless, it is still quite remarkable that the same protein is able to inhibit different caspases in fundamentally different manners.

One point pertinent to XIAP's ability to inhibit caspases does however deserve further consideration. Caspase-9 is activated downstream of mitochondrial membrane permeabilisation by cytochrome c that is released from the inter-mitochondrial membrane space by the action of the pro-apoptotic Bax/Bak molecules (Fig. 1b). It is widely accepted that permeabilisation of the outer mitochondrial membrane will cause death regardless of whether caspase-9 becomes activated or not. Thus it would seem that caspase-9 activation is not required for cell death. If this is the case, then it is hard to see how XIAP can prevent cell death, although it might delay the appearance of an apoptotic phenotype. Indeed, as many authors from the early period, including ourselves, looked at the ability of XIAP to "inhibit" cell death at a particular, and usually early, time point, rather than long-term cell survival, they may well have interpreted a slower death incorrectly as inhibition of cell death. And yet, XIAP binds to the activated, processed form of caspase-9 at the apoptosome indicating that caspase-9 has already been substantially activated by cytochrome c release from the inter-mitochondrial membrane space.

2.3 Different Hypotheses to Explain What XIAP Is Doing

So what is XIAP doing? Is it inhibiting cell death or is it modifying the way a cell dies? Both are possibilities. Because caspases have the ability to rapidly amplify their activity in a feedforward amplification loop, whereby initiator caspases activate downstream effector caspases which can in turn cleave upstream initiator caspases, a cell can potentially die a caspase mediated death very rapidly in the absence of mitochondrial disruption. In this scenario, XIAP may exist to deactivate low levels of accidentally activated caspase-9 or caspase-3 before they do serious damage. This would be the "*safety-catch*" hypothesis (Ditzel and Meier 2002; Fig. 1c). An extension of this hypothesis is the proposal that XIAP exists to dispose of inappropriately located proteins. Frequently proteins are processed on entry into the mitochondria and this generates an unusual, in a cytoplasmic context, alanine at the N-terminus of the protein. IAPs, via their BIR domains, are good at binding to such N-terminal Alanine proteins and this binding may activate their E3 ligase activity, thus allowing IAPs to bind to and destroy inappropriately activated (in the case of caspases) or inappropriately localised proteins that have escaped from the mitochondria into the cytoplasm. This could be considered as the "*policeman hypothesis*", (Ditzel and Meier 2002). Alternatively, there maybe non-lethal situations that we are not as yet aware of where the limited caspase activity is required and this is controlled by XIAP, or the "*regulator hypothesis*", (Fig. 1c). Certainly there are convincing experiments in *Drosophila* showing that caspase activity plays a non-apoptotic signalling role (Miura 2011; Orme et al. 2016). Finally, although it would have to be accepted strangely, it might be that XIAP prevents full caspase activation, preventing or delaying an apoptotic cell death (in the correct phenotypical sense), but failing to prevent cell death. Such activity might promote an inflammatory response that would not normally occur because the apoptotic phenotype helps the body dispose of dead cells in an anti-inflammatory manner (White et al. 2014; Rongvaux et al. 2014). This would then be the "*strange death*" hypothesis. This is perhaps the least likely hypothesis because, as we shall see, permeabilised mitochondria release many IAP antagonists, which could most likely overwhelm any XIAP inhibition and it is unlikely that all, or even most, of these antagonists would be missing in any particular cell (Verhagen et al. 2007). Thus we would hypothesise that there are always sufficient IAP antagonists to overcome XIAPs caspase inhibitory activity. Following this line of thought, it could be that XIAP serves as a safety catch but if it is not removed from activated caspases the outcome would not be surviving cells but rather a non-apoptotic cell death, which as we shall discuss in more detail in the necroptosis section, is inflammatory and potentially dangerous. Thus the *strange death* outcome could be considered as an unwanted drawback of a necessary *safety catch* function of XIAP and not really a function and that this drawback is negated by IAP antagonists. As a final thought on this topic it seems that the bacterium *Shigella flexneri* is able to manipulate its host cell to release Smac from the inter-mitochondrial membrane space without promoting apoptosis but antagonising the signalling function of XIAP (Andree et al. 2014).

One aspect of apoptotic cell death that is not often discussed is that in vivo apoptotic cells are extremely difficult to find, even though millions of cell deaths are occurring every minute (Silke and Johnstone 2016). This is because they are usually rapidly engulfed even before the plasma membrane becomes permeabilised. It is usually only when this process is completely overwhelmed by massive numbers of apoptotic cells that it becomes possible to detect them e.g. Rickard et al. 2014. This rapid removal in healthy tissues is most likely driven by caspase activity because caspases generate the phosphatidylserine signal exposed on dying cells that is detected by professional disposal cells such as macrophages (Segawa and Nagata 2015). All the details of what happens if this rapid caspase-dependent cell death and engulfment does not occur are not entirely clear but there is sufficient evidence to suggest that in several scenarios the result is increased inflammatory signalling (White et al. 2014; Rongvaux et al. 2014). Thus rapid, caspase-driven cell death might be as important in vivo as whether a cell dies or not and it is in such a situation that persistent anti-caspase activity of XIAP might be important. If we consider the *Apaf-1*, *caspase-9* and *caspase-3* knock-out mice, on a mixed 129/Sv C57BL/6 genetic background, these mice all suffer from a developmental exencephaly where the cerebrum extrudes beyond the uncompleted skull. At the time of these discoveries, it was assumed that *Apaf1*$^{-/-}$, *casp3*$^{-/-}$ or *casp9*$^{-/-}$ neurons failed to die as was developmentally appropriate and thereby prevented closure of the neural tube (Zheng et al. 1999). However, consistent with the idea that their activity is not required for cell death there are not increased numbers of neurons in *Apaf1*$^{-/-}$, *casp9*$^{-/-}$ and *casp3*$^{-/-}$ mice (Oppenheim et al. 2001; Nonomura et al. 2013). Rather the phenotype appears to arise because of a persistence of a very small number of undead cells in the neural ridge and a failure of the neural tube to close (Nonomura et al. 2013; Yamaguchi and Miura 2015; Oppenheim et al. 2001). As we would predict from our previous discussion, a number of cells in the neural ridge of the *Apaf1*$^{-/-}$ embryos were dying a non-apoptotic death, however many were described by the authors as "undead". Whether these undead cells were dying a slower non-apoptotic death or were somehow "zombie cells" is not clear to us but fits broadly with our idea that rapid apoptotic cell death that can be quickly cleared from the tissue is the primary function of the Apaf-1.caspase-9.caspase-3 cascade.

In theory if XIAP anti-caspase activity was not blocked, it might be able to similarly delay cell death of these neural ridge cells. At least one mechanism to prevent such an outcome is IAP antagonists. Such antagonists were first identified and recognised as such in *Drosophila* (Chen et al. 1996; White et al. 1994; Hay et al. 1995; Bergmann et al. 1998) and shown to bind to IAPs (Vucic et al. 1997). The first mammalian example Smac/DIABLO was identified a few years later (Du et al. 2000; Verhagen et al. 2000). Interestingly the *Drosophila* IAP antagonists are regulated transcriptionally while most, but not all, of the mammalian examples so far identified are secreted within the mitochondria and only released upon mitochondrial damage (Suzuki et al. 2001; Hegde et al. 2002; Martins et al. 2002; Verhagen et al. 2002; Hegde et al. 2003). One major mechanism for how IAP antagonists work was first determined in the mammalian counterparts. In essence, IAP antagonists bind to the same BIR domain and in more or less identical manner to the way that caspases bind. The neo N-terminus generated upon import into the inter-mitochondrial membrane space (or in the case of caspases by caspase cleavage) fits into the groove of the BIRs and thus competes directly for binding of IAPs with caspases. This N-terminus has been called an IAP binding motif (IBM) and minimally 4 amino acids are sufficient for the IBM/BIR interaction (Chai et al. 2000; Wu et al. 2000; Silke et al. 2000; Srinivasula et al. 2000). This IBM/BIR interaction discovery prompted the rapid development of small molecule Smac-mimetic peptide like drugs that were able to bind to IAPs and promote release of activated caspases (Flygare and Fairbrother 2010; Sun et al. 2008).

In summary, any hypothesis to explain the anti-caspase function of XIAP must account for the facts that it binds to activated caspases which should normally occur only following mitochondrial membrane permeabilisation, that cells with permeabilised mitochondria will die without activation of caspases and that nevertheless there are abundant mechanisms to antagonise XIAP and prevent it from inhibiting activated caspases. A primary difficulty for all the hypotheses that we have proposed is that XIAP deficient mice, in the lab, appear to have very little wrong with them, so if it does function as a policeman or a regulator it cannot be an essential activity in the life of a lab animal (Harlin et al. 2001; Olayioye et al. 2005). Furthermore while XIAP deficient humans do have significant problems (Rigaud et al. 2006), it would seem that these problems are usually unrelated to the ability of XIAP to inhibit apoptotic cell death and rather due to the ability of XIAP to regulate NOD signalling (Fig. 2b), and possibly the inflammasome (Fig. 3), which we shall discuss shortly (Krieg et al. 2009; Vince et al. 2012; Damgaard et al. 2012; Latour and Aguilar 2015; Silke and Vaux 2015).

3 IAPs Regulate Signalling and the Intersection with Cell Death

While XIAP has a ubiquitin E3 ligase role in regulating signalling from the innate immune receptors NOD1 and NOD2 this moves beyond our cell death focus. Therefore we will return to cIAPs at this point where the general role of IAPs in ubiquitin-regulated signalling was first discovered. Nevertheless cIAP regulation of TNFR1 signalling is analogous to XIAP regulation of NOD signalling as can be seen in Fig. 2, and the interested reader is referred to (Latour and Aguilar 2015; Hrdinka et al. 2016; Kupka et al. 2016b). While cIAPs were first discovered binding to TNFR2 they were soon after shown to bind to TNFR1 (Shu et al. 1996). In a strange twist of fate the discovery of the roles of cIAPs in these TNFR1 complexes relied heavily on the use of Smac-mimetics that were originally conceived and designed to inhibit XIAP. Given the foregoing discussion in the XIAP section, it was expected when these drugs were developed that they might enhance an apoptotic cell death when combined with a bona fide apoptotic stimulus, but would not have been expected to promote cell death on their own (Fulda et al. 2002;

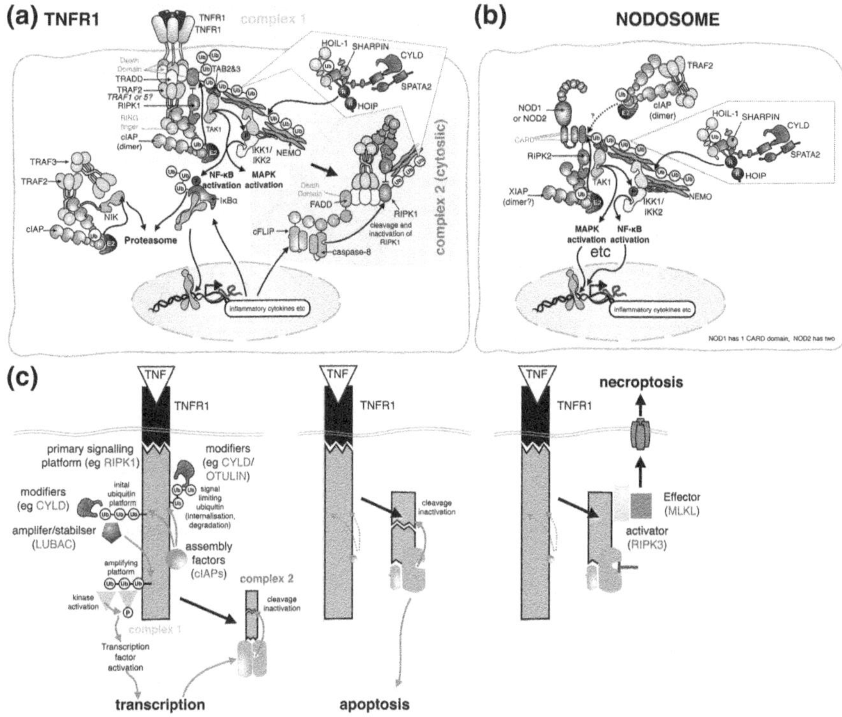

Fig. 2 **a** A schematic representation of the role that cIAPs play in regulating TNFR1 signalling. Ub; ubiquitin. Most players discussed or referenced in the text. **b** A schematic representation of XIAP regulation of NOD signalling highlighting similarities with cIAP regulation of TNF signalling. **c** Highly simplified schematic summarising information in (a) and how cIAPs inhibit apoptotic and necroptotic cell death

Li et al. 2004). Intriguingly however some of these compounds were able to promote cell death on their own, and this finding was one of the significant contributions to understanding the signalling role of cIAPs (Petersen et al. 2007; Gaither et al. 2007; Varfolomeev et al. 2007; Vince et al. 2007).

Together these papers, using different approaches, showed that Smac-mimetics did three amazing things. First, they promoted the rapid degradation of cIAPs, but not XIAP (Gaither et al. 2007; Varfolomeev et al. 2007; Vince et al. 2007). Second, they were able, in some cells, to promote production of TNF (Petersen et al. 2007; Gaither et al. 2007; Varfolomeev et al. 2007; Vince et al. 2007) and third that they sensitised cells to TNF-induced death (Vince et al. 2007). In a particularly conceptually important work at the time it was shown that cells could be sensitised to TRAIL-induced death by expression of a cytoplasmic form of Smac or delivery into cells of a Smac like peptide (Fulda et al. 2002). Subsequently it was shown that a Smac-mimetic could do the same, even with TNF (Li et al. 2004) but these experiments did not distinguish between the different IAPs. Nevertheless this newer

Hs NLRP3		
Position	Length	Domain
1-93	93	Pyrin
220-536	317	NACHT
742-762	21	LRR 1
771-792	22	LRR 2
799-819	21	LRR 3
828-849	22	LRR 4
856-876	21	LRR 5
885-906	22	LRR 6
913-933	21	LRR 7
942-963	22	LRR 8
970-991	22	LRR 9

Hs ASC		
Position	Length	Domain
1-91	91	Pyrin
107-195	89	CARD

Hs caspase-1		
Position	Length	Domain
1-91	91	CARD
120-297	178	p20
298-316	19	pro-peptide
317-404	88	p10

Hs Gasdermin D		
Position	Length	Domain
1-275	275	N-terminal
276-484	209	C-terminal

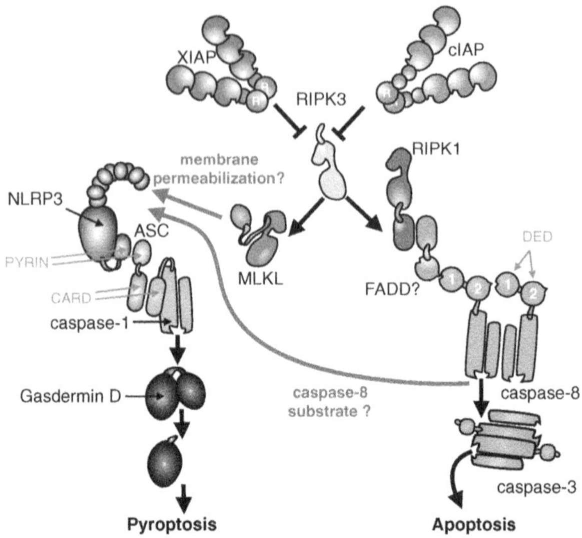

Fig. 3 A schematic summarising current knowledge on how IAPs regulate pyroptosis and inflammasome signalling. Boxes contain amino acid positions for human versions of the proteins

result was intriguing because while TRAIL kills many cancer cells (Wiley et al. 1995), TNF is not usually cytotoxic. This is widely considered to be because TNF drives NF-κB activation which results in production of cFLIP, an inactive caspase-8 homolog that dimerises with caspase-8 and affects its ability to oligomerise and activate (Micheau et al. 2001; Feoktistova et al. 2011; Dickens et al. 2012; Hughes et al. 2016). *cIap1*$^{-/-}$ MEFs like p65 *RelA*$^{-/-}$ MEFs were sensitive to TNF, like Smac-mimetic treated cells, suggesting that the activity of the different Smac-mimetics was on target and mimicked the loss of NF-κB activation (Vince et al. 2007). This supposition was backed up when it was shown that Smac-mimetic treatment or loss of cIAPs lead to a reduction of ubiquitylation of RIPK1, a marker for TNF activation of NF-κB, and reduced activation of NF-κB (Varfolomeev et al. 2008; Mahoney et al. 2008; Bertrand et al. 2009; Haas et al. 2009; Moulin et al. 2012). The picture that emerges from multiple lines of research is that the RING E3 ligase activity of cIAPs is required to ubiquitylate components within the TNFR1, which recruits the Linear Ubiquitin Assembly Complex (LUBAC), which helps drive activation of NF-κB and MAP Kinases (Haas et al. 2009; Feltham et al. 2010; Gerlach et al. 2011; Bertrand et al. 2011; Dynek et al. 2010; Blackwell et al. 2013). Perhaps surprisingly we have not been able to find published experiments showing that cFLIP expression levels are decreased in cIAP deficient cells which leaves open the possibility that cIAPs and LUBAC help limit TNF-induced death in other ways than by limiting production of cFLIP. Indeed one of the ways that cIAPs could limit a TNF-induced death signal is by ubiquitylating RIPK1 for destruction and preventing it from recruiting and activating caspase-8 via FADD or by otherwise regulating its activity (Bertrand et al. 2008; Ting and Bertrand 2016).

Given the specific signalling roles of different ubiquitin chains (Komander and Rape 2012), we would predict that the type of ubiquitin chain generated by cIAPs would help provide clues to their role. However, it seems that cIAPs are quite promiscuous and have the ability to generate many different types of ubiquitin chain including K11, K48 and K63 (Dynek et al. 2010; Blankenship et al. 2009) and even neddylate (Broemer et al. 2010). The simplest interpretation of a large amount of data, in broad outlines, is that cIAPs generate K63 ubiquitin chains that recruit TAB2/TAB3/TAK1 and LUBAC to the TNFR1 complex. LUBAC in turn generates linear ubiquitin chains that recruit NEMO/IKK1/IKK2 complexes with high affinity, and TAK1 is able to phosphorylate and activate IKK2 and MAPKs. However cIAPs may also generate K48 or K11 ubiquitin chains that promote proteasomal degradation (Fig. 2a, c). NEMO presence helps stabilise the complex and may do so by limiting the possibility of degradative ubiquitin chains. However it has also been proposed that cIAP generated K11 chains can recruit NEMO (Dynek et al. 2010). Of course if ubiquitin chains can cause particular types of signalling then their removal will counteract this and hence ubiquitin chains are heavily regulated by deubiquitylating enzymes called DUBs. And thus NEMO might limit DUB access to substrates. This important part of the signalling pathway is we feel beyond the scope of this chapter, but the interested reader is referred to (Kupka et al. 2016a; Wagner et al. 2016).

The evidence that cIAPs regulate the formation of RIPK1 containing complex 2 downstream of TNFR1 or other death receptor signalling is strong. In a detailed analysis of the CD95/Fas and downstream caspase-8 containing complex (complex 2), Geserick et al. showed nicely that Smac-mimetic treatment increased recruitment of RIPK1 into both of these complexes and that this recruitment could be reduced by ectopic expression of $cFLIP_L$ but not $cFLIP_S$ (Geserick et al. 2009). This indicates that cFLIP levels might not only determine whether caspase-8 can become activated to kill cells but might also work on another level by reducing complex 2 formation and thereby limiting caspase-8 activation in the first place. This idea that cIAPs limit the formation of a RIPK1 containing complex 2 was taken further by showing that "Ripoptosome" formation was provoked by a range of different stimuli if IAPs were inhibited (Feoktistova et al. 2011; Tenev et al. 2011) although Tenev et al. showed that combined loss of cIAP1, cIAP2 and XIAP induced the more Ripoptosome formation than loss of cIAP1, cIAP2 alone (Fig. 2c).

With reference to this last result, it is noteworthy that while cIAP antagonism or deletion in immortalised or cancer cell lines suffices to sensitise to TNF killing, in primary innate immune cells the picture is more complicated. Thus in bone marrow derived macrophages, dendritic cells or neutrophils, XIAP deletion can sensitise to LPS or TNF killing, while wild type cells remain resistant (Vince et al. 2012; Yabal et al. 2014; Lawlor et al. 2015; Wicki et al. 2016). This death is either apoptotic, as defined by caspase-8, caspase-3 and PARP cleavage in LPS-stimulated *Xiap*$^{-/-}$ cells (Vince et al. 2012; Yabal et al. 2014), or upon caspase-inhibition switches to RIPK3-MLKL mediated necroptotic killing (Lawlor et al. 2015). However contrary to most of our discussion, targeting cIAPs with cIAP-specific Smac-mimetics, or myeloid cell deletion of cIAP1 and cIAP2, does not dramatically sensitise macrophages to TNF- or LPS-induced apoptotic or necroptotic cell death, (Lawlor et al. 2015). Conceivably, loss of XIAP inhibition of effector caspase activity might play a role in sensitising macrophages to LPS or TNF killing, although the robust caspase-8 processing observed in LPS-stimulated *Xiap*$^{-/-}$ cells argues for XIAP acting upstream of caspase activation. Although they run counter to many other examples where cIAPs play the essential role in preventing TNF-induced death, these differences in innate immune cell compared with other cell types are not unprecedented. For example, TRADD plays an essential role in TNF-induced activation of NF-κB in fibroblast cells, but not macrophages (Ermolaeva et al. 2008; Pobezinskaya et al. 2008; Chen et al. 2008).

4 IAPs and Necroptosis

In addition to being useful for uncovering the role of cIAPs in inhibiting apoptosis, Smac-mimetics have also been extremely widely used to promote necroptosis. Indeed a standard necroptotic stimulus entails using TNF to activate TNFR1 signalling, a Smac-mimetic to deplete cIAPs and thereby promote TNF-induced cell death and a caspase inhibitor to ensure that caspase-8 is blocked leading to necroptosis. A similar treatment can also be used with other death ligands of the TNF superfamily, such as TRAIL and FasL. Smac-mimetics probably have two actions that facilitate formation of a necrosome, first they prevent TNF-induced, cIAP mediated, transcriptional upregulation of pro-survival targets like cFLIP and second they prevent modification of RIPK1 allowing levels of this to build up and potentially oligomerise and activate RIPK3, nevertheless the most common outcome following TNF + Smac-mimetic treatment is apoptosis, in order to drive cells down a necroptotic signalling pathway it is still usually necessary to inhibit caspase-8 activity. While older papers use Z-VAD-FMK and Q-VD-OPh to great effect, we recently found that the clinical caspase-8 inhibitor emricasan was effective at much lower concentrations because it is a good inhibitor of both the caspase-8 homodimer and the caspase-8/cFLIP$_L$ heterodimer (Brumatti et al. 2016).

Loss of both cIAP1 and cIAP2 causes early embryonic lethality at Embyonic Day 10.5 (E10.5) and this lethality can be delayed by concomitant loss of either TNFR1, RIPK1 or RIPK3 (Moulin et al. 2012). *Tnfr1* loss gives the greatest protection, till birth, while *Ripk1* and *Ripk3* loss allow embryogenesis to proceed for only an extra two days (Moulin et al. 2012). It should be noted that a question has been raised about whether the *cIap1*$^{-/-}$ mice reported in this work might also have a defect in cIAP2 translation, however as we are mostly concerned here with the *cIap1*$^{-/-}$*cIap2*$^{-/-}$ mice this is not particularly concerning (Moulin et al. 2015) The early lethal phenotype of the *cIap1*$^{-/-}$*cIap2*$^{-/-}$ mice is remarkably similar to that of the *Casp8*$^{-/-}$ and *Fadd*$^{-/-}$ mice which can also be prevented by loss of

Ripk1, *Ripk3* or *Tnfr1* (Oberst et al. 2011; Kaiser et al. 2011; Zhang et al. 2011). This phenotype is also highly similar to that of mice deficient in HOIP1, an essential LUBAC component (Peltzer et al. 2014), and here too the early lethality is prevented by loss of *Tnfr1* (Peltzer et al. 2014).

Volumes have been written on the necroptotic cell death pathway recently and as IAP inhibition is usually used to promote it the role of IAPs in this process is also dealt with in many reviews. Therefore if the reader is interested in a more detailed discussion on this topic, we recommend (Pasparakis and Vandenabeele 2015; Silke and Vaux 2014). One interesting twist to this tale is that while Smac-mimetics were developed to kill cancer cells they have been used to uncover a role of IAPs in inhibiting necroptosis. This in turn has led to the idea that Smac-mimetics can be used to promote necroptosis which might be an effective way to kill apoptosis resistant cancers (Steinhart et al. 2013; Steinwascher et al. 2015; Brumatti et al. 2016; McComb et al. 2016).

5 IAPs and Pyroptosis

The inflammatory caspases, caspase-1 and caspase-11, can induce a lytic pro-inflammatory cell death termed pyroptosis. The activation of these caspases required to execute this pathway is dependent on the formation of large protein complexes termed inflammasomes. Caspase-11 forms the so-called non-canonical inflammasome and senses cytoplasmic LPS derived from Gram-negative bacteria (Kayagaki et al. 2011, 2013; Hagar et al. 2013) or host-derived oxidised phospholipids (Zanoni et al. 2016). LPS, but not oxidised phospholipid, activated caspase-11 subsequently cleaves Gasdermin D at Asp275 to unleash an N-terminal fragment that oligomerises and forms pores in the plasma membrane, resulting in pyroptotic killing (Shi et al. 2015; Kayagaki et al. 2015; Chen et al. 2016; Ding et al. 2016; He et al. 2015).

In contrast to caspase-11 activation by LPS, a variety of canonical inflammasome sensor proteins exist that can detect many pathogen, environmental and host danger molecules to induce caspase-1, but not caspase-11, activation. These include several members of the NOD-Like Receptor (NLR) family, including NLRP1, NLRP3 and NLRC4, that when mutated, drive auto-inflammatory diseases through excessive inflammasome formation and resulting caspase-1 activation (Masters et al. 2012; Menu and Vince 2011). Like caspase-11, active caspase-1 also cleaves and activates Gasdermin D to induce a rapid pyroptotic cell death. To date, most inflammasome-associated diseases appear to result from an increased ability of caspase-1 to cleave and thereby activate the pro-inflammatory cytokine, Interleukin-1β (IL-1β), although some evidence is emerging implicating caspase-1 pyroptosis in host–pathogen responses (Miao et al. 2010).

Research implicating XIAP in the regulation of inflammation in humans emerged when it was shown that patients deficient in XIAP (X-linked lymphoproliferative syndrome type 2, XLP-2), are predisposed to inflammatory bowel disease and hemophagocytic lymphohistiocytosis (HLH) that is associated with viral infections, such as (Rigaud et al. 2006; Latour and Aguilar 2015). A correlation with the severity of HLH and sustained, significantly elevated, inflammasome-associated IL-18 levels, but not other cytokines such as TNF and IL-6, has been observed (Wada et al. 2013). XIAP deficiency has also been associated with clinical features resembling those found in people harbouring auto-activating NLRP3 inflammasome mutations (Christiansen et al. 2016).

Recent studies using Smac-mimetic compounds and IAP deficient mice have documented how IAPs might negatively regulate NLRP3 inflammasome signalling and consequent caspase-1, IL-1β and IL-18 activation in innate immune cells (Fig. 3; Vince et al. 2012; Yabal et al. 2014; Lawlor et al. 2015; Wicki et al. 2016). As is likely to occur in humans, a deficiency in XIAP expression predisposes mouse macrophages, dendritic cells and neutrophils to spontaneous Toll-like Receptor (TLR)-induced precursor IL-1β cleavage to the active, highly pro-inflammatory, secreted fragment. While the loss of cIAPs alone does not mimic XIAP deletion, the targeting of both cIAPs and XIAP with Smac-mimetics, or their combined genetic deletion in myeloid cells, greatly exacerbates TLR- or TNF-induced IL-1β activation, which is observed in vitro and in vivo, and results in a systemic TNF-driven inflammatory disease with arthritic features (Wong et al. 2014; Lawlor et al. 2015).

Although the key target of IAPs, and in particularly XIAP, required to repress TLR or TNF-driven IL-1β activation remains unclear, it is likely to occur at or upstream of RIPK3 signalling, as the deletion of RIPK3 or RIPK3 and caspase-8 ablates IL-1β activation following IAP loss (Lawlor et al. 2015). This RIPK3-caspase-8 dependent activation of IL-1β occurs through both activation of NLRP3-caspase-1, and the direct cleavage of IL-1β by caspase-8 (Vince et al. 2012; Lawlor et al. 2015; Conos et al. 2016). Of note, the inhibition of caspase-8, or its deletion, can also result in spontaneous TLR-induced NLRP3 signalling via the necroptotic effector MLKL (Kang et al. 2013; Lawlor et al. 2015; Kang et al. 2015; Gaidt et al. 2016). How caspase-8 and MLKL activate NLRP3 remains unclear, although this could conceivably occur through membrane damage and potassium efflux, which is commonly believed to trigger NLRP3 oligomerisation and caspase-1 recruitment (Muñoz-Planillo et al. 2013). However, this notion does not align with reports indicating that RIPK3-MLKL induced activation of NLRP3 (Kang et al. 2013), as well as caspase-8 induced NLRP3 (Gaidt et al. 2016), can occur in the absence of death in murine dendritic cells and human monocytes, respectively.

Because RIPK3-MLKL and caspase-8 can activate NLRP3, the resulting caspase-1 processing and activation may also, in theory, result in Gasdermin D cleavage and pyroptotic cell death. However, caspase-8 induced apoptosis, or MLKL-dependent necroptosis, may suffice to kill cells following the activation of these molecules, even when NLRP3 or caspase-1 is absent, although experimental evidence testing this hypothesis yet to be reported.

6 Conclusion

The study of mammalian IAPs has advanced far beyond their original identification where all that was known was that they had domains with no known function and that they could block apoptosis. However, we think that there are still many important issues that remain unresolved and unclear. We have a lot of molecular detail for how XIAP inhibit caspases 3 and 9 and how this inhibition is relieved by IAP antagonists yet in some ways this only accentuates the fact that the "reason" for such activity, according to current understanding and as discussed above, is far from obvious. It would also be helpful to discern whether there is a common thread between the signalling aspect of XIAP where its ubiquitin ligase activity is important and regulation of caspases where if such activity plays a role it has been difficult to show, although see Schile et al. (2008). In other words, what is the biological logic that links these two different functions of XIAP?

cIAP 1 and 2 clearly have a signalling role that is intricately entwined with inhibition of caspase-8 activation. Despite earlier claims that these IAPs inhibit caspase-3 and caspase-7, the consensus seems to be that they bind but do not inhibit these caspases (Eckelman and Salvesen 2006). They do inhibit activation of caspase-8 but it has been shown that they do not inhibit activated caspase-8 (Roy et al. 1997). In our opinion, it would be remarkable if XIAP and cIAPs had very different ways of working against caspases suggesting that cIAPs work in death receptor induced signalling complexes in an analogous way to XIAP at the apoptosome but evidence for this is non-existent.

Another particularly important point that might help us understand the role of IAPs is why there are two cIAPs? cIAP1 and 2 are extremely closely linked at the genetic level (Kenneth 2012; Moulin et al. 2012) implying a recent genetic duplication event (Cao et al. 2008), yet they appear to have different characteristics at the biochemical level (Feltham et al. 2011). We favour the view that cIAP1 is the dominant partner in the relationship-based primarily on the fact that in mice that we generated, combined loss of XIAP and cIAP2 was tolerated and the mice were viable whereas combined loss of cIAP1 and cIAP2 or cIAP1 and XIAP resulted in embryonic lethality (Moulin et al. 2012). However this result is not as clear cut as it seems and Heard et al. have generated viable $Xiap^{-/-}.cIap1^{-/-}$ mice (Moulin et al. 2015; Heard et al. 2015). Others seem to favour the idea that cIAP1 and cIAP2 activity is indistinguishable (Bertrand et al. 2008; Mahoney et al. 2008). One issue is that cIAP2 normally seems to be expressed at much lower levels than cIAP1 and in mice is difficult to reliably detect; developing good tools to look at this issue is becoming an increasing priority.

Another huge question is the precise role of cIAPs in regulating the formation of a cytosolic RIPK1 containing cell death inducing complex, variously called complex 2 or the Ripoptosome. While we have presented a simplified picture for the role of cIAPs earlier in this chapter, how the different ubiquitin chains that can be generated by cIAPs at the TNFR1 complex are formed and in what way this is regulated remains unclear. Furthermore, how lack of cIAPs promotes formation of the complex 2 and in what way normal TNFR1 signalling prevents formation is still, we believe, unclear though many possibilities exist.

Some of these questions will undoubtedly have clinical and translational relevance but often times it is hard to predict whether discoveries will or will not. In situations where they are not immediately translatable, we should however remember those early IAP pioneers who made such an important contribution to our knowledge providing many of us with the happy chance to translate into the clinic (Clem 2015).

References

Andree M, Seeger JM, Schüll S, Coutelle O et al (2014) BID-dependent release of mitochondrial SMAC dampens XIAP-mediated immunity against Shigella. EMBO J 33:2171–2187

Bergmann A, Agapite J, McCall K, Steller H (1998) The Drosophila Gene Hid is a direct molecular target of ras-dependent survival signaling. Cell 95:331–341

Bertrand MJ, Milutinovic S, Dickson KM, Ho WC et al (2008) cIAP1 and cIAP2 facilitate cancer cell survival by functioning as E3 ligases that promote RIP1 ubiquitination. Mol Cell 30:689–700

Bertrand MJ, Doiron K, Labbé K, Korneluk RG et al (2009) Cellular inhibitors of apoptosis cIAP1 and cIAP2 are required for innate immunity signaling by the pattern recognition receptors NOD1 and NOD2. Immunity 30(6):789–801

Bertrand MJ, Lippens S, Staes A, Gilbert B et al (2011) cIAP1/2 are direct E3 ligases conjugating diverse types of ubiquitin chains to receptor interacting proteins kinases 1 to 4 (RIP1-4). PLoS ONE 6:e22356

Bird FT (1964) On the development of insect polyhedrosis and granulosis virus particles. Can J Microbiol 10:49–52

Birnbaum MJ, Clem RJ, Miller LK (1994) An apoptosis inhibiting gene from a nuclear polyhedrosis virus encoding a polypeptide with cys/his sequence motif. J Virol 68:2521–2528

Blackwell K, Zhang L, Workman LM, Ting AT et al (2013) Two coordinated mechanisms underlie tumor necrosis factor alpha-induced immediate and delayed IκB kinase activation. Mol Cell Biol 33:1901–1915

Blankenship JW, Varfolomeev E, Goncharov T, Fedorova AV et al (2009) Ubiquitin binding modulates IAP antagonist-stimulated proteasomal degradation of c-IAP1 and c-IAP2. Biochem J 417:149–160

Boatright KM, Renatus M, Scott FL, Sperandio S et al (2003) A unified model for apical caspase activation. Mol Cell 11:529–541

Broemer M, Tenev T, Rigbolt KT, Hempel S et al (2010) Systematic in vivo RNAi analysis identifies IAPs as NEDD8-E3 ligases. Mol Cell 40:810–822

Brumatti G, Ma C, Lalaoui N, Nguyen NY et al (2016) The caspase-8 inhibitor emricasan combines with the SMAC mimetic birinapant to induce necroptosis and treat acute myeloid leukemia. Sci Transl Med 8:339ra69

Cao L, Wang Z, Yang X, Xie L, Yu L (2008) The evolution of BIR domain and its containing proteins. FEBS Lett 582:3817–3822

Chai J, Du C, Wu JW, Kyin S et al (2000) Structural and biochemical basis of apoptotic activation by Smac/DIABLO. Nature 406:855–862

Chai J, Shiozaki E, Srinivasula SM, Wu Q et al (2001) Structural basis of caspase-7 inhibition by XIAP. Cell 104:769–780

Chen P, Nordstrom W, Gish B, Abrams JM (1996) Grim, a novel cell death gene in Drosophila. Genes Dev 10:1773–1782

Chen NJ, Chio II, Lin WJ, Duncan G et al (2008) Beyond tumor necrosis factor receptor: TRADD signaling in toll-like receptors. Proc Natl Acad Sci USA 105:12429–12434

Chen X, He WT, Hu L, Li J et al (2016) Pyroptosis is driven by non-selective gasdermin-D pore and its morphology is different from MLKL channel-mediated necroptosis. Cell Res 26: 1007–1020

Christiansen M, Ammann S, Speckmann C, Mogensen TH (2016) XIAP deficiency and MEFV variants resulting in an autoinflammatory lymphoproliferative syndrome. BMJ Case Rep

Clem RJ (2015) Viral IAPs, then and now. Semin Cell Dev Biol 39:72–79

Conos SA, Lawlor KE, Vaux DL, Vince JE, Lindqvist LM (2016) Cell death is not essential for caspase-1-mediated interleukin-1β activation and secretion. Cell Death Differ 23:1827–1838

Crook NE, Clem RJ, Miller LK (1993) An apoptosis inhibiting baculovirus gene with a zinc finger like motif. J Virol 67:2168–2174

Damgaard RB, Nachbur U, Yabal M, Wong WW et al (2012) The Ubiquitin Ligase XIAP Recruits LUBAC for NOD2 signaling in inflammation and innate immunity. Mol Cell 46:746–758

Dannappel M, Vlantis K, Kumari S, Polykratis A et al (2014) RIPK1 maintains epithelial homeostasis by inhibiting apoptosis and necroptosis. Nature 513:90–94

Deveraux QL, Takahashi R, Salvesen GS, Reed JC (1997) X-linked IAP is a direct inhibitor of cell-death proteases. Nature 388:300–304

Deveraux QL, Roy N, Stennicke HR, Van Arsdale T et al (1998) IAPs block apoptotic events induced by caspase-8 and cytochrome c by direct inhibition of distinct caspases. EMBO J 17:2215–2223

Deveraux QL, Leo E, Stennicke HR, Welsh K et al (1999) Cleavage of human inhibitor of apoptosis protein XIAP results in fragments with distinct specificities for caspases. EMBO J 18:5242–5251

Dickens LS, Boyd RS, Jukes-Jones R, Hughes MA et al (2012) A death effector domain chain DISC model reveals a crucial role for caspase-8 chain assembly in mediating apoptotic cell death. Mol Cell 47:291–305

Ding J, Wang K, Liu W, She Y et al (2016) Pore-forming activity and structural autoinhibition of the gasdermin family. Nature 535:111–116

Ditzel M, Meier P (2002) IAP degradation: decisive blow or altruistic sacrifice? Trends Cell Biol 12:449–452

Du C, Fang M, Li Y, Li L, Wang X (2000) Smac, a mitochondrial protein that promotes cytochrome c–dependent caspase activation by eliminating IAP inhibition. Cell 102:33–42

Duckett CS, Nava VE, Gedrich RW, Clem RJ et al (1996) A conserved family of cellular genes related to the baculovirus iap gene and encoding apoptosis inhibitors. EMBO J 15:2685–2694

Dynek JN, Goncharov T, Dueber EC, Fedorova AV et al (2010) c-IAP1 and UbcH5 promote K11-linked polyubiquitination of RIP1 in TNF signalling. EMBO J 29:4198–4209

Eckelman BP, Salvesen GS (2006) The human anti-apoptotic proteins cIAP1 and cIAP2 bind but do not inhibit caspases. J Biol Chem 281:3254–3260

Ermolaeva MA, Michallet MC, Papadopoulou N, Utermöhlen O et al (2008) Function of TRADD in tumor necrosis factor receptor 1 signaling and in TRIF-dependent inflammatory responses. Nat Immunol 9:1037–1046

Etemadi N, Chopin M, Anderton H, Tanzer MC et al (2015) TRAF2 regulates TNF and NF-κB signalling to suppress apoptosis and skin inflammation independently of Sphingosine kinase-1. Elife 4:e10592

Feltham R, Moulin M, Vince JE, Mace PD et al (2010) Tumor necrosis factor (TNF) signaling, but not TWEAK (TNF-like weak inducer of apoptosis)-triggered cIAP1 (cellular inhibitor of apoptosis protein 1) degradation, requires cIAP1 RING dimerization and E2 binding. J Biol Chem 285:17525–17536

Feltham R, Bettjeman B, Budhidarmo R, Mace PD et al (2011) Smac mimetics activate the E3 ligase activity of cIAP1 protein by promoting RING domain dimerization. J Biol Chem 286:17015–17028

Feoktistova M, Geserick P, Kellert B, Dimitrova DP et al (2011) cIAPs block ripoptosome formation, a RIP1/Caspase-8 containing intracellular cell death complex differentially regulated by cFLIP isoforms. Mol Cell 43:449–463

Fisher AJ, Cruz W, Zoog SJ, Schneider CL, Friesen PD (1999) Crystal structure of baculovirus P35: role of a novel reactive site loop in apoptotic caspase inhibition. EMBO J 18:2031–2039

Flygare JA, Fairbrother WJ (2010) Small-molecule pan-IAP antagonists: a patent review. Expert Opin Ther Pat 20:251–267

Fulda S, Wick W, Weller M, Debatin KM (2002) Smac agonists sensitize for Apo2L/TRAIL- or anticancer drug-induced apoptosis and induce regression of malignant glioma in vivo. Nat Med 8:808–815

Gaidt MM, Ebert TS, Chauhan D, Schmidt T et al (2016) Human monocytes engage an alternative inflammasome pathway. Immunity 44:833–846

Gaither A, Porter D, Yao Y, Borawski J et al (2007) A Smac mimetic rescue screen reveals roles for inhibitor of apoptosis proteins in tumor necrosis factor-alpha signaling. Cancer Res 67:11493–11498

Gentle IE, Silke J (2011) New perspectives in TNF-R1-induced NF-κB signaling. Adv Exp Med Biol, pp. 79–88 United States

Gerlach B, Cordier SM, Schmukle AC, Emmerich CH et al (2011) Linear ubiquitination prevents inflammation and regulates immune signalling. Nature 471:591–596

Geserick P, Hupe M, Moulin M, Wong WW et al (2009) Cellular IAPs inhibit a cryptic CD95-induced cell death by limiting RIP1 kinase recruitment. J Cell Biol 187:1037–1054

Haas TL, Emmerich CH, Gerlach B, Schmukle AC et al (2009) Recruitment of the linear ubiquitin chain assembly complex stabilizes the TNF-R1 signaling complex and is required for TNF-mediated gene induction. Mol Cell 36:831–844

Hagar JA, Powell DA, Aachoui Y, Ernst RK, Miao EA (2013) Cytoplasmic LPS activates caspase-11: implications in TLR4-independent endotoxic shock. Science 341:1250–1253

Harlin H, Reffey SB, Duckett CS, Lindsten T, Thompson CB (2001) Characterization of XIAP-deficient mice. Mol Cell Biol 21:3604–3608

Hay BA, Wassarman DA, Rubin GM (1995) Drosophila homologs of baculovirus inhibitor of apoptosis proteins function to block cell death. Cell 83:1253–1262

He WT, Wan H, Hu L, Chen P et al (2015) Gasdermin D is an executor of pyroptosis and required for interleukin-1β secretion. Cell Res 25:1285–1298

Heard KN, Bertrand MJ, Barker PA (2015) cIAP2 supports viability of mice lacking cIAP1 and XIAP. EMBO J 34:2393–2395

Hegde R, Srinivasula SM, Zhang Z, Wassell R et al (2002) Identification of Omi/HtrA2 as a mitochondrial apoptotic serine protease that disrupts inhibitor of apoptosis protein-caspase interaction. J Biol Chem 277:432–438

Hegde R, Srinivasula SM, Datta P, Madesh M et al (2003) The polypeptide chain-releasing factor GSPT1/eRF3 is proteolytically processed into an IAP-binding protein. J Biol Chem 278:38699–38706

Hrdinka M, Fiil BK, Zucca M, Leske D et al (2016) CYLD limits Lys63- and Met1-linked ubiquitin at receptor complexes to regulate innate immune signaling. Cell Rep 14:2846–2858

Huang Y, Park YC, Rich RL, Segal D et al (2001) Structural basis of caspase inhibition by XIAP. Differential roles of the linker versus the BIR domain. Cell 104:781–790

Hughes MA, Powley IR, Jukes-Jones R, Horn S et al (2016) Co-operative and hierarchical binding of c-FLIP and Caspase-8: a unified model defines how c-FLIP isoforms differentially control cell fate. Mol Cell 61:834–849

Inceoglu AB, Kamita SG, Hammock BD (2006) Genetically modified baculoviruses: a historical overview and future outlook. Adv Virus Res 68:323–360

Joazeiro CA, Wing SS, Huang H, Leverson JD et al (1999) The tyrosine kinase negative regulator c-Cbl as a RING-type, E2- dependent ubiquitin-protein ligase. Science 286:309–312

Kaiser WJ, Upton JW, Long AB, Livingston-Rosanoff D et al (2011) RIP3 mediates the embryonic lethality of caspase-8-deficient mice. Nature 471:368–372

Kang TB, Yang SH, Toth B, Kovalenko A, Wallach D (2013) Caspase-8 blocks kinase RIPK3-mediated activation of the NLRP3 inflammasome. Immunity 38:27–40
Kang S, Fernandes-Alnemri T, Rogers C, Mayes L et al (2015) Caspase-8 scaffolding function and MLKL regulate NLRP3 inflammasome activation downstream of TLR3. Nat Commun 6:7515
Kayagaki N, Warming S, Lamkanfi M, Vande Walle L et al (2011) Non-canonical inflammasome activation targets caspase-11. Nature 479:117–121
Kayagaki N, Wong MT, Stowe IB, Ramani SR et al (2013) Noncanonical inflammasome activation by intracellular LPS independent of TLR4. Science 341:1246–1249
Kayagaki N, Stowe IB, Lee BL, O'Rourke K et al (2015) Caspase-11 cleaves gasdermin D for non-canonical inflammasome signalling. Nature 526:666–671
Kenneth NS, Younger JM, Hughes ED, Marcotte D, et al. (2012) An inactivating caspase 11 passenger mutation originating from the 129 murine strain in mice targeted for c-IAP1. Biochem J 443:355–359
Komander D, Rape M (2012) The ubiquitin code. Annu Rev Biochem 81:203–229
Krieg A, Correa RG, Garrison JB, Le Negrate G et al (2009) XIAP mediates NOD signaling via interaction with RIP2. Proc Natl Acad Sci USA 106:14524–14529
Kupka S, Reichert M, Draber P, Walczak H (2016a) Formation and removal of poly-ubiquitin chains in the regulation of tumor necrosis factor-induced gene activation and cell death. FEBS J 283:2626–2639
Kupka S, De Miguel D, Draber P, Martino L et al (2016b) SPATA2-mediated binding of CYLD to HOIP enables CYLD recruitment to signaling complexes. Cell Rep 16:2271–2280
Latour S, Aguilar C (2015) XIAP deficiency syndrome in humans. Semin Cell Dev Biol 39:115–123
Lawlor KE, Khan N, Mildenhall A, Gerlic M et al (2015) RIPK3 promotes cell death and NLRP3 inflammasome activation in the absence of MLKL. Nat Commun 6:6282
Lee SY, Reichlin A, Santana A, Sokol KA et al (1997) Traf2 is essential for jnk but not nf-kappa-b activation and regulates lymphocyte proliferation and survival. Immunity 7:703–713
Li P, Nijhawan D, Budihardjo I, Srinivasula SM et al (1997) Cytochrome c and dATP-dependent formation of Apaf-1/caspase-9 complex initiates an apoptotic protease cascade. Cell 91:479–489
Li L, Thomas RM, Suzuki H, De Brabander JK et al (2004) A small molecule Smac mimic potentiates TRAIL- and TNFα-mediated cell death. Science 305:1471–1474
Liston P, Roy N, Tamai K, Lefebvre C et al (1996) Suppression of apoptosis in mammalian cells by naip and a related family of iap genes. Nature 379:349–353
Mace PD, Smits C, Vaux DL, Silke J, Day CL (2010) Asymmetric recruitment of cIAPs by TRAF2. J Mol Biol 400:8–15
Mahoney DJ, Cheung HH, Mrad RL, Plenchette S et al (2008) Both cIAP1 and cIAP2 regulate TNFα-mediated NF-κB activation. Proc Natl Acad Sci USA 105:11778–11783
Martins LM, Iaccarino I, Tenev T, Gschmeissner S et al (2002) The serine protease Omi/HtrA2 regulates apoptosis by binding XIAP through a Reaper-like motif. J Biol Chem 277:439–444
Masters SL, Gerlic M, Metcalf D, Preston S et al (2012) NLRP1 inflammasome activation induces pyroptosis of hematopoietic progenitor cells. Immunity 37:1009–1023
McComb S, Aguadé-Gorgorió J, Harder L, Marovca B et al (2016) Activation of concurrent apoptosis and necroptosis by SMAC mimetics for the treatment of refractory and relapsed ALL. Sci Transl Med 8:339ra70
Menu P, Vince JE (2011) The NLRP3 inflammasome in health and disease: the good, the bad and the ugly. Clin Exp Immunol 166:1–15
Miao EA, Leaf IA, Treuting PM, Mao DP et al (2010) Caspase-1-induced pyroptosis is an innate immune effector mechanism against intracellular bacteria. Nat Immunol 11:1136–1142
Micheau O, Lens S, Gaide O, Alevizopoulos K, Tschopp J (2001) NF-κB signals induce the expression of c-FLIP. Mol Cell Biol 21:5299–5305
Miura M (2011) Apoptotic and non-apoptotic caspase functions in neural development. Neurochem Res 36:1253–1260

Moulin M, Anderton H, Voss AK, Thomas T et al (2012) IAPs limit activation of RIP kinases by TNF receptor 1 during development. EMBO J 31:1679–1691
Moulin M, Voss AK, Thomas T, Wong WW et al (2015) Response to Heard et al. EMBO J 34:2396–2397
Muñoz-Planillo R, Kuffa P, Martínez-Colón G, Smith BL et al (2013) K+ efflux is the common trigger of NLRP3 inflammasome activation by bacterial toxins and particulate matter. Immunity 38:1142–1153
Nonomura K, Yamaguchi Y, Hamachi M, Koike M et al (2013) Local apoptosis modulates early mammalian brain development through the elimination of morphogen-producing cells. Dev Cell 27:621–634
Oberst A, Dillon CP, Weinlich R, McCormick LL et al (2011) Catalytic activity of the caspase-8-FLIP(L) complex inhibits RIPK3-dependent necrosis. Nature 471:363–367
Olayioye MA, Kaufmann H, Pakusch M, Vaux DL et al (2005) XIAP-deficiency leads to delayed lobuloalveolar development in the mammary gland. Cell Death Differ 12:87–90
Oppenheim RW, Flavell RA, Vinsant S, Prevette D et al (2001) Programmed cell death of developing mammalian neurons after genetic deletion of caspases. J Neurosci 21:4752–4760
Orme MH, Liccardi G, Moderau N, Feltham R et al (2016) The unconventional myosin CRINKLED and its mammalian orthologue MYO7A regulate caspases in their signalling roles. Nat Commun 7:10972
Park YC, Burkitt V, Villa AR, Tong L, Wu H (1999) Structural basis for self-association and receptor recognition of human TRAF2. Nature 398:533–538
Park YC, Ye H, Hsia C, Segal D et al (2000) A novel mechanism of TRAF signaling revealed by structural and functional analyses of the TRADD-TRAF2 interaction. Cell 101:777–787
Pasparakis M, Vandenabeele P (2015) Necroptosis and its role in inflammation. Nature 517:311–320
Peltzer N, Rieser E, Taraborrelli L, Draber P et al (2014) HOIP deficiency causes embryonic lethality by aberrant TNFR1-mediated endothelial cell death. Cell Rep 9:153–165
Petersen SL, Wang L, Yalcin-Chin A, Li L et al (2007) Autocrine TNFalpha signaling renders human cancer cells susceptible to Smac-mimetic-induced apoptosis. Cancer Cell 12:445–456
Pobezinskaya YL, Kim YS, Choksi S, Morgan MJ et al (2008) The function of TRADD in signaling through tumor necrosis factor receptor 1 and TRIF-dependent Toll-like receptors. Nat Immunol 9:1047–1054
Rickard JA, O'Donnell JA, Evans JM, Lalaoui N et al (2014) RIPK1 regulates RIPK3-MLKL-driven systemic inflammation and emergency hematopoiesis. Cell 157:1175–1188
Riedl SJ, Renatus M, Schwarzenbacher R, Zhou Q et al (2001) Structural basis for the inhibition of caspase-3 by XIAP. Cell 104:791–800
Rigaud S, Fondanèche MC, Lambert N, Pasquier B et al (2006) XIAP deficiency in humans causes an X-linked lymphoproliferative syndrome. Nature 444:110–114
Rongvaux A, Jackson R, Harman CC, Li T et al (2014) Apoptotic caspases prevent the induction of type I interferons by mitochondrial DNA. Cell 159:1563–1577
Rothe M, Pan MG, Henzel WJ, Ayres TM, Goeddel DV (1995) The TNF-R2-TRAF signaling complex contains two novel proteins related to baculoviral-inhibitor of apoptosis proteins. Cell 83:1243–1252
Roy N, Deveraux QL, Takahashi R, Salvesen GS, Reed JC (1997) The c-iap-1 and c-iap-2 proteins are direct inhibitors of specific caspases. EMBO J 16:6914–6925
Samuel T, Welsh K, Lober T, Togo SH et al (2006) Distinct BIR domains of cIAP1 mediate binding to and ubiquitination of TRAF2 and Smac. J Biol Chem 281:1080–1090
Schile AJ, García-Fernández M, Steller H (2008) Regulation of apoptosis by XIAP ubiquitin-ligase activity. Genes Dev 22:2256–2266
Segawa K, Nagata S (2015) An apoptotic 'Eat Me' signal: phosphatidylserine exposure. Trends Cell Biol 25:639–650
Shi J, Zhao Y, Wang K, Shi X et al (2015) Cleavage of GSDMD by inflammatory caspases determines pyroptotic cell death. Nature 526:660–665

Shu HB, Takeuchi M, Goeddel DV (1996) The tumor necrosis factor receptor 2 signal transducers traf2 and c-iap1 are components of the tumor necrosis factor receptor 1 signaling complex. Proc Natl Acad Sci USA 93:13973–13978

Silke J, Brink R (2010) Regulation of TNFRSF and innate immune signalling complexes by TRAFs and cIAPs. Cell Death Differ 17:35–45

Silke J, Johnstone RW (2016) In the midst of life-cell death: what is it, what is it good for, and how to study it. Cold Spring Harb Protoc 2016:pdb.top070508

Silke J, Meier P (2013) Inhibitor of apoptosis (IAP) proteins-modulators of cell death and inflammation. Cold Spring Harb Perspect Biol 5:2 (United States)

Silke J, Vaux D (2014) IAPs and necroptotic cell death. In: Shen HM, Vandenabeele P (eds) Necrotic cell death. Springer, New York, pp 57–77

Silke J, Vaux DL (2015) IAP gene deletion and conditional knockout models. Semin Cell Dev Biol 39:97–105

Silke J, Verhagen AM, Ekert PG, Vaux DL (2000) Sequence as well as functional similarity for DIABLO/smac and grim, reaper and hid? Cell Death Differ 7:1275

Silke J, Ekert PG, Day CL, Hawkins CJ et al (2001) Direct inhibition of caspase 3 is dispensable for the anti-apoptotic activity of XIAP. EMBO J 20:3114–3123

Srinivasula SM, Datta P, Fan XJ, Fernandes-Alnemri T et al (2000) Molecular determinants of the caspase-promoting activity of smac/DIABLO and its role in the death receptor pathway. J Biol Chem 275:36152–36157

Steinhart L, Belz K, Fulda S (2013) Smac mimetic and demethylating agents synergistically trigger cell death in acute myeloid leukemia cells and overcome apoptosis resistance by inducing necroptosis. Cell Death Dis 4:e802

Steinwascher S, Nugues AL, Schoeneberger H, Fulda S (2015) Identification of a novel synergistic induction of cell death by Smac mimetic and HDAC inhibitors in acute myeloid leukemia cells. Cancer Lett 366:32–43

Sun C, Cai M, Gunasekera AH, Meadows RP et al (1999) NMR structure and mutagenesis of the inhibitor-of-apoptosis protein XIAP. Nature 401:818–821

Sun H, Nikolovska-Coleska Z, Yang CY, Qian D et al (2008) Design of small-molecule peptidic and nonpeptidic Smac mimetics. Acc Chem Res 41:1264–1277

Suzuki Y, Imai Y, Nakayama H, Takahashi K et al (2001) A serine protease, HtrA2, is released from the mitochondria and interacts with XIAP, inducing cell death. Mol Cell 8:613–621

Tada K, Okazaki T, Sakon S, Kobarai T et al (2001) Critical roles of TRAF2 and TRAF5 in tumor necrosis factor-induced NF-κ B activation and protection from cell death. J Biol Chem 276:36530–36534

Takahashi R, Deveraux Q, Tamm I, Welsh K et al (1998) A single BIR domain of XIAP sufficient for inhibiting caspases. J Biol Chem 273:7787–7790

Takahashi N, Vereecke L, Bertrand MJM, Duprez L et al (2014) RIPK1 ensures intestinal homeostasis by protecting the epithelium against apoptosis. Nature 513:95–99

Tenev T, Bianchi K, Darding M, Broemer M et al (2011) The Ripoptosome, a signaling platform that assembles in response to genotoxic stress and loss of IAPs. Mol Cell 43:432–448

Ting AT, Bertrand MJ (2016) More to life than NF-κB in TNFR1 signaling. Trends Immunol 37:535–545

Uren AG, Pakusch M, Hawkins CJ, Puls KL, Vaux DL (1996) Cloning and expression of apoptosis inhibitory protein homologs that function to inhibit apoptosis and/or bind TRAFs. Proc Natl Acad Sci USA 93:4974–4978

Varfolomeev E, Wayson SM, Dixit VM, Fairbrother WJ, Vucic D (2006) The inhibitor of apoptosis protein fusion c-IAP2.MALT1 stimulates NF-κB activation independently of TRAF1 and TRAF2. J Biol Chem 281:29022–29029

Varfolomeev E, Blankenship JW, Wayson SM, Fedorova AV et al (2007) IAP antagonists induce autoubiquitination of c-IAPs, NF-κB activation, and TNFα-dependent apoptosis. Cell 131:669–681

Varfolomeev E, Goncharov T, Fedorova AV, Dynek JN et al (2008) c-IAP1 and c-IAP2 are critical mediators of tumor necrosis factor alpha (TNFα)-induced NF-κB activation. J Biol Chem 283:24295–24299

Varfolomeev E, Goncharov T, Maecker H, Zobel K et al (2012) Cellular inhibitors of apoptosis are global regulators of NF-κB and MAPK activation by members of the TNF family of receptors. Sci Signal 5:ra22

Verhagen AM, Ekert PG, Pakusch M, Silke J et al (2000) Identification of DIABLO, a mammalian protein that promotes apoptosis by binding to and antagonizing IAP proteins. Cell 102:42–53

Verhagen AM, Silke J, Ekert PG, Pakusch M et al (2002) HtrA2 promotes cell death through its serine protease activity and its ability to antagonize inhibitor of apoptosis proteins. J Biol Chem 277:445–454

Verhagen AM, Kratina TK, Hawkins CJ, Silke J et al (2007) Identification of mammalian mitochondrial proteins that interact with IAPs via N-terminal IAP binding motifs. Cell Death Diff 14:348–357

Vince JE, Wong WW, Khan N, Feltham R et al (2007) IAP antagonists target cIAP1 to induce TNFα-dependent apoptosis. Cell 131:682–693

Vince JE, Pantaki D, Feltham R, Mace PD et al (2009) TRAF2 must bind to cellular inhibitors of apoptosis for tumor necrosis factor (TNF) to efficiently activate NF-{κ}B and to prevent TNF-induced apoptosis. J Biol Chem 284:35906–35915

Vince JE, Wong WW, Gentle I, Lawlor KE et al (2012) Inhibitor of apoptosis proteins limit RIP3 kinase-dependent interleukin-1 activation. Immunity 36:215–227

Vucic D, Kaiser WJ, Harvey AJ, Miller LK (1997) Inhibition of reaper-induced apoptosis by interaction with inhibitor of apoptosis proteins (IAPs). Proc Natl Acad Sci USA 94:10183–10188

Wada T, Muraoka M, Yokoyama T, Toma T et al (2013) Cytokine profiles in children with primary Epstein-Barr virus infection. Pediatr Blood Cancer 60:E46–E48

Wagner SA, Satpathy S, Beli P, Choudhary C (2016) SPATA2 links CYLD to the TNF-α receptor signaling complex and modulates the receptor signaling outcomes. EMBO J 35:1868–1884

White K, Grether ME, Abrams JM, Young L et al (1994) Genetic control of programmed cell death in Drosophila. Science 264:677–683

White MJ, McArthur K, Metcalf D, Lane RM et al (2014) Apoptotic caspases suppress mtDNA-induced STING-mediated type I IFN production. Cell 159:1549–1562

Wicki S, Gurzeler U, Wei-Lynn Wong W, Jost PJ et al (2016) Loss of XIAP facilitates switch to TNFα-induced necroptosis in mouse neutrophils. Cell Death Dis 7:e2422

Wiley SR, Schooley K, Smolak PJ, Din WS et al (1995) Identification and characterization of a new member of the TNF family that induces apoptosis. Immunity 3:673–682

Wong WW, Gentle IE, Nachbur U, Anderton H et al (2010) RIPK1 is not essential for TNFR1-induced activation of NF-κB. Cell Death Differ 17:482–487

Wong WW, Vince JE, Lalaoui N, Lawlor KE et al (2014) cIAPs and XIAP regulate myelopoiesis through cytokine production in an RIPK1- and RIPK3-dependent manner. Blood 123:2562–2572

Wu G, Chai JJ, Suber TL, Wu JW et al (2000) Structural basis of IAP recognition by Smac/DIABLO. Nature 408:1008–1012

Yabal M, Müller N, Adler H, Knies N et al (2014) XIAP restricts TNF- and RIP3-dependent cell death and inflammasome activation. Cell Rep 7:1796–1808

Yamaguchi Y, Miura M (2015) Programmed cell death in neurodevelopment. Dev Cell 32:478–490

Yang Y, Fang S, Jensen JP, Weissman AM, Ashwell JD (2000) Ubiquitin protein ligase activity of IAPs and their degradation in proteasomes in response to apoptotic stimuli. Science 288:874–877

Ye H, Park YC, Kreishman M, Kieff E, Wu H (1999) The structural basis for the recognition of diverse receptor sequences by TRAF2. Mol Cell 4:321–330

Yeh WC, Shahinian A, Speiser D, Kraunus J et al (1997) Early lethality, functional nf-kb activation, and increased sensitivity to TNF-induced cell death in TRAF2-deficient mice. Immunity 7:715–725

Zanoni I, Tan Y, Di Gioia M, Broggi A et al (2016) An endogenous caspase-11 ligand elicits interleukin-1 release from living dendritic cells. Science 352:1232–1236

Zhang H, Zhou X, McQuade T, Li J et al (2011) Functional complementation between FADD and RIP1 in embryos and lymphocytes. Nature 471:373–376

Zhang L, Blackwell K, Workman LM, Gibson-Corley KN et al (2016) TRAF2 exerts opposing effects on basal and TNFα-induced activation of the classic IKK complex in hematopoietic cells in mice. J Cell Sci 129:1455–1467

Zheng TS, Hunot S, Kuida K, Flavell RA (1999) Caspase knockouts: matters of life and death. Cell Death Differ 6:1043–1053

Zheng C, Kabaleeswaran V, Wang Y, Cheng G, Wu H (2010) Crystal structures of the TRAF2: cIAP2 and the TRAF1: TRAF2: cIAP2 complexes: affinity, specificity, and regulation. Mol Cell 38:101–113

Zhou Q, Krebs JF, Snipas SJ, Price A et al (1998) Interaction of the baculovirus anti-apoptotic protein p35 with caspases—specificity, kinetics, and characterization of the caspase/p35 complex. Biochemistry 37:10757–10765

Cellular FLICE-Inhibitory Protein Regulates Tissue Homeostasis

Hiroyasu Nakano, Xuehua Piao, Ryodai Shindo and Sachiko Komazawa-Sakon

Abstract Cellular FLICE-inhibitory protein (cFLIP) is structurally related to caspase-8, but lacks its protease activity. *Cflip* gene encodes several splicing variants including short form (cFLIPs) and long form ($cFLIP_L$). $cFLIP_L$ is composed of two death effector domains at the N terminus and a C-terminal caspase-like domain, and cFLIPs lacks the caspase-like domain. Our studies reveal that cFLIP plays a central role in NF-κB-dependent survival signals that control apoptosis and programmed necrosis. Germline deletion of *Cflip* results in embryonic lethality due to enhanced apoptosis and programmed necrosis; however, the combined deletion of the death-signaling regulators, *Fadd* and *Ripk3,* prevents embryonic lethality in *Cflip*-deficient mice. Moreover, tissue-specific deletion of *Cflip* reveals cFLIP as a crucial regulator that maintains tissue homeostasis of immune cells, hepatocytes, intestinal epithelial cells, and epidermal cells by preventing apoptosis and programmed necrosis.

Abbreviations

BHA	Butylated hydroxyanisole
Ciap	Cellular inhibitor of apoptosis
cFLIP	Cellular FLICE-inhibitory protein
CCCP	Carbonyl cyanide m-chlorophenylhydrazone
DED	Death effector domain
DPI	Diphenyleneiodonium
EHV2	Equine herpes virus-2
ERK	Extracellular signal-regulated kinase
FADD	Fas-associated protein with death domain
HHV-8	Human herpes virus-8

H. Nakano (✉) · X. Piao · R. Shindo · S. Komazawa-Sakon
Department of Biochemistry, Toho University School of Medicine, 5-21-16 Omori-Nishi, Ota-Ku, Tokyo 143-8540, Japan
e-mail: hiroyasu.nakano@med.toho-u.ac.jp

R. Shindo
Department of Immunology, Juntendo University Graduate School of Medicine, 2-1-1 Hongo, Bunkyo-Ku, Tokyo 113-8421, Japan

Current Topics in Microbiology and Immunology (2017) 403:119–142
DOI 10.1007/82_2015_448

Published Online: 08 July 2015

IKKβ	IκB kinase β
IEC	Intestinal epithelial cell
JNK	C-Jun N-terminal kinase
LUBAC	Linear ubiquitination chain assembly complex
MCV	Molluscum contagiosum virus
MLKL	Mixed lineage kinase domain-like
MKK7	Mitogen-activated protein (MAP) kinase kinase 7
MEK1	MAPK/ERK kinase 1
Nec-1	Necrostatin-1
NEMO	NF-κB essential modulator
NF-κB	Nuclear factor-κB
Nox	NADPH oxidase
RIPK	Receptor-interacting protein kinase
TAK1	TGF-β-activated kinase 1
TCA	Tricarboxylic acid cycle
TNF-α	Tumor necrosis factor-α
TRAIL	TNF-related apoptosis-inducing ligand
TRADD	TNF receptor-associated death domain
TRAF	TNF receptor-associated factor

Contents

1 Introduction

Apoptosis is the prototype of programmed cell death or regulated cell death and is executed by sequential activation of cysteine proteases (caspases) (Riedl and Salvesen 2007; Yuan 2006). Recent studies have revealed several alternative forms of programmed cell death including necroptosis, pyroptosis, and ferroptosis (Mocarski et al. 2014; Pasparakis and Vandenabeele 2015; Tait et al. 2014). Among them, receptor-interacting protein kinase 1 (RIPK1)-dependent necrosis is referred to necroptosis and is inhibited by necrostatin-1 (Nec-1), an inhibitor of RIPK1 (Degterev et al. 2005, 2008). Execution of the extrinsic apoptotic pathway is triggered by death receptors such as tumor necrosis factor receptor 1 (TNFR1), Fas, or TNF-related death-inducing ligand (TRAIL) receptor. TNF-α usually induces formation of a complex I that subsequently activates NF-κB, resulting in induction of inflammatory cytokine genes and cell survival. However, under certain conditions such as in the presence of protein synthesis inhibitors, TNF-α stimulation induces the formation of a complex IIa that is made up of TNF receptor-associated death domain (TRADD), Fas-associated protein with death domain (FADD), and caspase-8 (Fig. 1). When expression of cIAP1 and 2 is downregulated in the presence of IAP antagonists, TNF-α stimulation results in formation of a complex

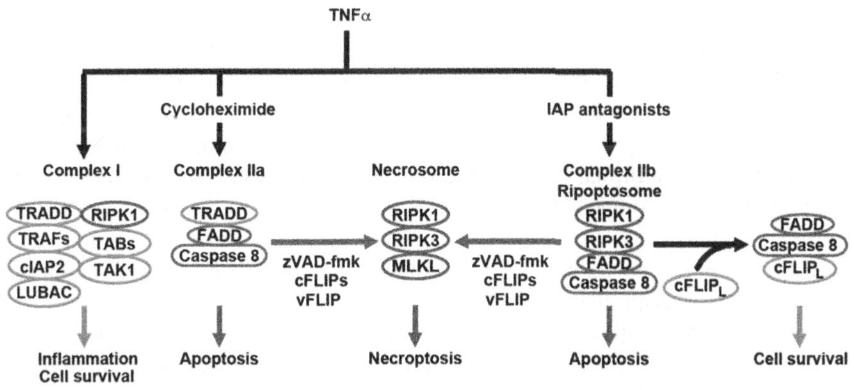

Fig. 1 Signaling pathways triggered by TNF-α. In many types of cells, TNF-α induces formation of a complex I that subsequently activates NF-κB, resulting in induction of inflammatory cytokine genes and cell survival under normal conditions. However, under certain conditions such as in the presence of a protein synthesis inhibitor, cycloheximide (CHX), TNF-α stimulation induces formation of a complex IIa that is made up of TRADD, FADD, and caspase-8. When expression of cIAP1 and 2 is downregulated in the presence of IAP antagonists, TNF-α stimulation results in the formation of a complex IIb, which is also referred to as the ripoptosome that is composed of FADD, RIPK1, RIPK3, and caspase-8. Moreover, when cFLIP$_L$ is recruited to the ripoptosome via interaction with caspase-8, the caspase-8/cFLIP$_L$ heterodimer exhibits a limited, but not full protease activity that subsequently cleaves and inactivates RIPK1, RIPK3, and CYLD. Therefore, the caspase-8/cFLIP$_L$ heterodimer blocks both apoptosis and necroptosis. In sharp contrast, caspase-8 activity is compromised in the presence of cFLIPs, vFLIP, or zVAD-fmk, and the complex IIa and complex IIb evolve to the necrosome that is composed of RIPK1, RIPK3, and MLKL. While the complex IIa and complex IIb promote apoptosis, the necrosome promotes necroptosis

IIb, which is also referred to as the ripoptosome that is composed of FADD, RIPK1, RIPK3, and caspase-8 (Mocarski et al. 2014; Pasparakis and Vandenabeele 2015; Tait et al. 2014). Notably, both the complex IIa and complex IIb evolve to form the necrosome under the conditions in which caspase-8 activity is suppressed in the presence of caspase inhibitors. While complex IIa and complex IIb promote apoptosis, the necrosome induces necroptosis. Accumulating studies have shown that cellular FLICE-inhibitory protein (cFLIP) plays a central role in suppression of death receptor-induced necroptosis as well as apoptosis (Budd et al. 2006; Silke and Strasser 2013). In this review, we will introduce recent progress in cell death signaling focusing on functions of cFLIP and also discuss the contribution of reactive oxygen species to necroptosis.

2 Structure and Function of cFLIP

Scanning expression databases for homologs of death effector domain (DED) involved in apoptosis, Thome et al. identified genes encoding DED in several viral genomes, viral FLIP (*vFLIP*). They also identified two cellular homologs of *vFLIP*, designated as short and long forms of cellular FLIP (*CFLIP*) (Irmler et al. 1997; Thome et al. 1997). Several groups also identified cFLIP, which is referred to as CASH, Casper, CLARP, MRIT, I-FLICE, or FLAME (Goltsev et al. 1997; Han et al. 1997; Hu et al. 1997; Inohara et al. 1997; Irmler et al. 1997; Shu et al. 1997; Srinivasula et al. 1997). Single *Cflip* gene in mammalian genome encodes $cFLIP_L$ and $cFLIP_S$ due to alternative splicing, although additional splicing variants have been reported (Golks et al. 2005). $cFLIP_L$ is composed of N-terminal two DEDs and C-terminal caspase-like domain lacking cysteine protease activity, whereas cFLIPs is composed of two DEDs (Fig. 2). Therefore, cFLIPs and $cFLIP_L$ are highly homologous to vFLIP and caspase-8, respectively. $cFLIP_L$ and cFLIPs have been reported to be cleaved by caspases or regulated by ubiquitylation and phosphorylation (Safa 2013).

Since the identification of *Cflip* gene, the debate continues over the role of cFLIP in promoting or inhibiting apoptosis in vitro. Several studies have shown that overexpression of $cFLIP_L$, but not cFLIPs, promotes apoptosis through activation of caspase-8 in a cell-type-dependent manner (Goltsev et al. 1997; Shu et al. 1997). Moreover, affinity of interaction between caspase-8 and $cFLIP_L$ is higher than caspase-8/caspase-8 homodimer, and caspase-like domain of $cFLIP_L$ induces partial cleavage of caspase-8, resulting in efficient activation of caspase-8 (Yu et al. 2009). In sharp contrast, other studies have shown that overexpression of $cFLIP_L$ inhibits death receptor-induced apoptosis (Irmler et al. 1997). Conversely, knockdown of *Cflip* by siRNAs or germline deletion of *Cflip* results in an increase in susceptibility to apoptosis (Yeh et al. 2000). How do we reconcile these apparently inconsistent results? One of the plausible explanations would be that expression levels of $cFLIP_L$ need to be kept in very narrow ranges in cells to suppress apoptosis under

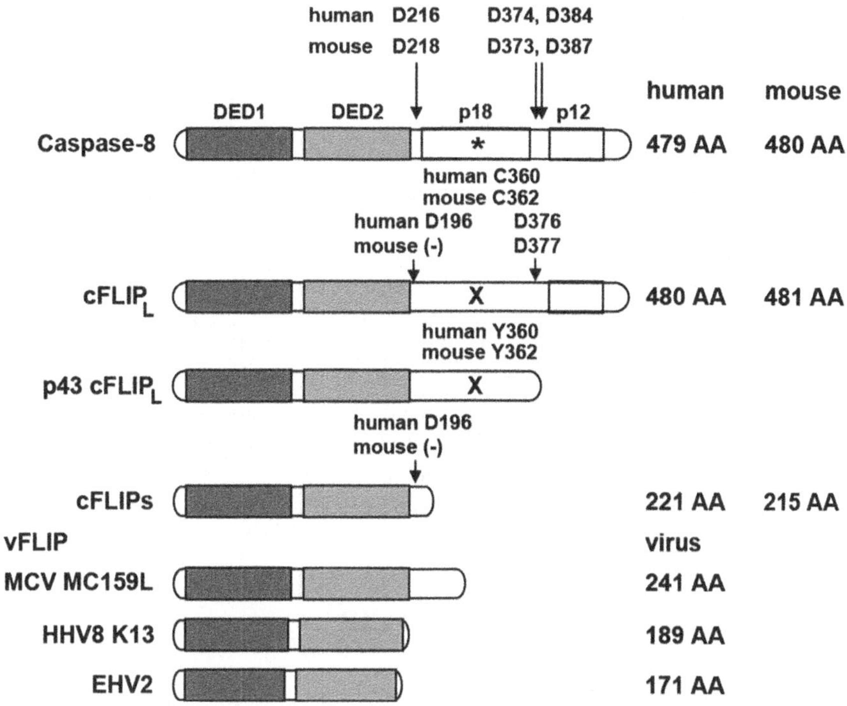

Fig. 2 Structures of caspase-8 and cFLIP isoforms. Caspase-8 is composed of N-terminal two DEDs and C-terminal caspase domain. Oligomerization of caspase-8 results in the generation of p18 and p12 fragments by trans- and auto-processing of caspase-8. A complex composed of the p18–p12 heterotetramer is an active form of caspase-8. On the other hand, $cFLIP_L$ is composed of N-terminal two DEDs and C-terminal caspase-like domain. Cysteine residues at an enzyme active site corresponding to caspase-8 are replaced by tyrosine (human Y360 and mouse Y362); therefore, $cFLIP_L$ does not exhibit protease activity. $cFLIP_L$ is partially cleaved by caspase-8, resulting in a p43 and p12 $cFLIP_L$ fragments. cFLIPs and vFLIP are composed of N-terminal two DEDs. Cleaved aspartic residues of each protein are indicated. Notably, asparatic acid residue corresponding to human D196 does not exist in murine cFLIP. DED; death effector domain. Cysteine residues at an enzyme active site of human and mouse caspase-8 represent C360 and C362, respectively. *MCV* molluscum contagiosum virus; *EHV-2* equine herpes virus-2; *HHV-8* human herpes virus-8. Numbers indicate amino acids corresponding to the indicated proteins

steady-state conditions. Upregulation or downregulation of $cFLIP_L$ above or below a certain level might promote apoptosis.

In sharp contrast to $cFLIP_L$, vFLIP and cFLIPs suppress death receptor-induced apoptosis (Irmler et al. 1997; Thome et al. 1999). Moreover, the crystal structure of vFLIP (MC159) reveals that MC159 disrupts the formation of the Fas signaling complex containing FADD and caspase-8 by forming the FADD/MC159 and the caspase-8/MC159 heterodimers (Yang et al. 2005). Therefore, cFLIPs is considered to be a genuine inhibitor of apoptosis; however, this concept is challenged by a recent study (Oberst et al. 2011).

3 cFLIP Regulates Necroptosis

In addition to regulation of apoptosis, cFLIP also regulates necroptosis. RIPK1 plays a central role in the formation of the complex IIb under conditions in which cellular inhibitor of apoptosis (IAP) is depleted (Feoktistova et al. 2011; Tenev et al. 2011). When $cFLIP_L$ is recruited to the ripoptosome via interaction with caspase-8, the caspase-8/$cFLIP_L$ heterodimer cleaves RIPK1 and RIPK3, resulting in disruption of the ripoptosome (Fig. 1) (Feoktistova et al. 2011; Oberst et al. 2011). The caspase-8/$cFLIP_L$ heterodimer also cleaves and inactivates a deubiquitinase named CYLD that promotes necroptosis (O'Donnell et al. 2011). Therefore, $cFLIP_L$ blocks necroptosis as well as apoptosis. In contrast, when cFLIPs is recruited to the ripoptosome, the caspase-8/cFLIPs heterodimer cannot cleave RIPK1 or RIPK3, rather promotes formation of the necrosome and necroptosis (Fig. 1) (Feoktistova et al. 2011; Oberst et al. 2011). However, these results were solely obtained in vitro experiments, and it is crucial to test whether such phenomenon is observed in mice overexpressing cFLIPs.

4 Regulation of Expression of cFLIP

$cFLIP_L$ is ubiquitously expressed in various tissues, whereas the expression of cFLIPs is rather restricted to several tumor cells and activated lymphocytes (Budd et al. 2006). While many transcription factors have been reported to regulate *CFLIP* (Safa 2013), NF-κB is crucial for the regulation of *CFLIP* transcription in several cell types (Kreuz et al. 2001; Micheau et al. 2001). In addition to regulation of *CFLIP* at mRNA levels, cFLIP is a very unstable protein and rapidly undergoes degradation by the proteasome-dependent pathway (Palacios et al. 2006). We and others have found that TNF-α induces rapid degradation of $cFLIP_L$ in NF-κB-deficient fibroblasts, whereas TNF-α does not upregulate $cFLIP_L$ even in wild-type fibroblasts at protein levels (Chang et al. 2006; Nakajima et al. 2006). This indicates that activation of NF-κB might prevent degradation of $cFLIP_L$ by the proteasome. Chang et al. found that an E3 ligase named ITCH is responsible for degradation of $cFLIP_L$ (Chang et al. 2006). Although prolonged c-Jun N-terminal kinase (JNK) activation promotes apoptosis under certain conditions (De Smaele et al. 2001; Gao et al. 2004; Tang et al. 2001), the target(s) of the JNK pathway to promote apoptosis is not fully understood. Taken that ITCH is activated by JNK and degrades $cFLIP_L$ (Gao et al. 2004), ITCH might be an attractive candidate to missing link between the JNK pathway and apoptosis. However, *Itch*-deficient mice exhibit a complex phenotype including severe interstitial pneumonia, stomach and skin inflammation, and enhanced T helper (Th)2 type responses (Perry et al. 1998; Venuprasad et al. 2006). Moreover, a later study showed that proteasome-dependent degradation of $cFLIP_L$ depends on JNK, but not ITCH (Sanchez-Perez et al. 2010).

The issue remains unclear as whether ITCH is the sole ubiquitin ligase regulating $cFLIP_L$, or alternatively, a ligase other than ITCH may be primarily responsible for degradation of $cFLIP_L$.

5 Other Signaling Pathways Regulated by cFLIP

5.1 cFLIP and the NF-κB Pathway

Several studies have shown that caspase-8 cleaves $cFLIP_L$, resulting in the generation of a processed p43 fragment of $cFLIP_L$ (Kataoka et al. 2000; Kavuri et al. 2011; Koenig et al. 2014). This fragment subsequently interacts with TRAF2 and activates NF-κB. Since caspase-8 activates NF-κB in some cell lines (Imamura et al. 2004), $cFLIP_L$-dependent NF-κB activation explains, at least in part, the mechanisms underlying caspase-8-dependent NF-κB activation. However, it is unclear whether caspase-8 and $cFLIP_L$-dependent NF-κB activation might have biological significance under physiological conditions, since TNF-α-induced NF-κB activation does not appear to be impaired in *Cflip*-deficient fibroblasts (Nakajima et al. 2006).

5.2 cFLIP and the MAPK Pathway

In addition to the effects of cFLIP on the NF-κB pathway, cFLIP also regulates the mitogen-activated protein kinase (MAPK) pathways. We previously reported that overexpression of $cFLIP_L$, but not cFLIPs, suppresses JNK and extracellular signal-regulated kinase (ERK), but not p38MAPK through directly binding to MAP kinase kinase 7 (MKK7) and MAPK/ERK kinase 1/2 (MEK1/2), respectively (Nakajima et al. 2006). Moreover, we found that TNF-α induces caspase-dependent sustained JNK and ERK activation in several tumor cell lines including HeLa and HCT116 cells with knockdown of *CFLIP* (Nakajima et al. 2008). These results suggest that $cFLIP_L$ directly and/or indirectly suppresses the JNK and ERK pathways. On the other hand, Grambihler et al. reported that $cFLIP_L$ inhibits p38MAPK activation, thereby suppressing bile acid-induced apoptosis (Grambihler et al. 2003). Taken that sustained JNK activation promotes cell death under certain conditions, these studies have revealed a novel anti-apoptotic function of $cFLIP_L$ other than direct binding and suppressing caspase-8. In sharp contrast, other groups reported that overexpression of $cFLIP_L$ activates the ERK pathway (Kataoka et al. 2000; Koenig et al. 2014). The discrepancy between these results is currently unclear, but it might come from different cells used for experiments, or different experimental conditions including overexpression of $cFLIP_L$ or siRNA-mediated depletion of *CFLIP*. Further study will be required to establish a consensus model to explain these inconsistent results.

5.3 cFLIP and the Wnt Pathway

cFLIP also activates the Wnt signaling pathway. Overexpression of $cFLIP_L$ suppresses β-catenin ubiquitinylation, resulting in nuclear translocation of β-catenin and induction of the β-catenin-dependent genes in both *Xenopus* embryos and mammalian cells (Naito et al. 2004). In addition, overexpression of vFLIP similarly enhances the Wnt signaling pathway (Nakagiri et al. 2005). These results reveal the multifaceted roles mediated by cFLIP; however, it remains unclear whether these functions are relevant to physiological functions of cFLIP.

6 cFLIP Plays a Crucial Role in Tissue Homeostasis

6.1 Germline Deletion of *Cflip* Results in Embryonic Lethality Due to Enhanced Apoptosis and Necroptosis

Genetic deletion of *Cflip* results in embryonic lethality due to a defect in the vascular development of yolk sac (Yeh et al. 2000). Intriguingly, *Fadd*- and/or *Caspase-8*-deficient mice also exhibit similar phenotype (Sakamaki et al. 2002; Varfolomeev et al. 1998; Zhang et al. 1998). Although cFLIP, FADD, and caspase-8 are involved in both execution and suppression of the extrinsic apoptotic pathways, *Tnfr1*, *Fas*, or *Trail* deficiency does not phenocopy *Cflip*-, *Fadd*-, or *Caspase-8*-deficient mice (Adachi et al. 1995; Cretney et al. 2002; Pfeffer et al. 1993). These results suggest that cFLIP, FADD, and caspase-8 might have unique function(s) other than execution or suppression of apoptosis induced by death receptors. While the mechanisms by which the deletion of *Fadd, Cflip*, and *Caspase-8* causes embryonic lethality remain obscure, recent studies reported that deletion of *Ripk1* or *Ripk3* rescues embryonic lethality of *Fadd*- and *Caspase-8*-deficient mice by preventing necroptosis (Kaiser et al. 2011; Oberst et al. 2011; Zhang et al. 2011). Moreover, combined deletion of *Fadd* and *Ripk3* is required for preventing lethal phenotype of *Cflip*-deficient mice (Dillon et al. 2012). These results suggest that cFLIP, FADD, and caspase-8 play crucial roles in normal development by preventing RIPK1- and RIPK3-dependent necroptosis. Furthermore, several groups including us have reported tissue-specific *Cflip*-deficient mice, which revealed an essential role for cFLIP in tissue homeostasis.

6.2 Roles for cFLIP in Immune Cells

T cell-, B cell-, macrophage-, and eosinophil-specific *Cflip*-deficient mice have been generated (Table 1). Zhang et al. reported that numbers of mature T cells are reduced in T cell-specific *Cflip*-deficient mice (Zhang and He 2005). Moreover,

Table 1 Summary of tissue-specific *Cflip*-deficient mice

	Cre driver lines spontaneous phenotypes		Types of cell death in other features		References
Immune cells					
T cells	*Lck-Cre*	Lack of mature T cells and impaired proliferation of T cells by TCR	Apoptosis	An increase in susceptibility to anti-Fas Ab-induced apoptosis	Zhang and He (2005)
B cells	*CD19-Cre*	Decrease in mature B cells and impaired proliferation of B cells by TLRs and LPS	Apoptosis	An increase in susceptibility to anti-Fas Ab-induced apoptosis	Zhang et al. (2009b)
Macrophages and Eosinophils	*LysM-Cre and Rosa26-CreER*	Loss of marginal zone macrophages, elevation of peripheral blood neutrophils, splenomegaly and enhanced apoptosis of eosinophils	Apoptosis		Gordy et al. (2011, 2014)
IECs	*Villin-Cre*	Perinatal lethality and Severe colitis	Apoptosis and necroptosis	Partial rescue by crossing with *Tnfr1-/-* mice	Piao et al. (2012)
	Villin-Cre	Perinatal lethality and Severe colitis			Wittkopf et al. (2013)
	Villin-CreER	Severe colitis		Failure to ameliorate colitis by anti-TNF-α antibody and RIPK3 independent	Wittkopf et al. (2013)
	Rosa26-CreER	Severe colitis		Amelioration by anti-TNF-α antibody and RIPK3 independent	Weinlich et al. (2013)
Epidermis	*Rosa26-CreER*	Severe dermatitis after tamoxifen injection	apoptosis	Amelioration by anti-TNF-α antibody	Weinlich et al. (2013)
	K14-CreER	Severe dermatitis after tamoxifen injection	apoptosis	Amelioration by anti-TNF-α antibody	Panayotova-Dimitrova et al. (2013)
	K14-Cre	Embryonic lethal		Undetermined	Panayotova-Dimitrova et al. (2013)

(continued)

Table 1 (continued)

	Cre driver lines spontaneous phenotypes		Types of cell death in other features		References
Hepatocytes	*Albumin-Cre*	Normal liver function	Apoptosis	An increase in susceptibility to anti-Fas Ab *and* ConA-induced hepatitis	Scahttenberg et al. (2011)
	Albumin-Cre	Normal liver function	Apoptosis	An increase in susceptibility to anti-Fas Ab-induced hepatitis	Piao et al. (2012)
	Alfp-Cre	Perinatal lethality and severe hepatitis	Apoptosis and necroptosis	Failure to rescue lethal phenotype by crossing with *Tnfr1-/-* mice	Piao et al. (2012)
	Mx1-Cre	Severe hepatitis after PolyI:C injection	Apoptosis	Amelioration by anti-TNF-α, anti-FasL, and anti-TRAIL antibodies	Piao et al. (2012)

Cflip-deficient T cells exhibit an impaired proliferation upon TCR stimulation (Chau et al. 2005; Zhang and He 2005). Similarly, numbers of B cells are reduced in peripheral tissues of B cell-specific *Cflip*-deficient mice (Zhang et al. 2009b). *Cflip*-deficient B cells increase a susceptibility to Fas-induced apoptosis, but exhibit an aberrant upregulation of costimulatory molecules and activation markers. The mechanism underlying hyperactivation of *Cflip*-deficient B cells is currently unknown. On the other hand, marginal zone and bone marrow stromal macrophages disappear in *CflipF/F;LysM-Cre* mice (Gordy et al. 2011). Delayed clearance of apoptotic neutrophils by macrophages results in granulocyte colony-stimulating factor (G-CSF)-dependent severe neutrophilia, splenomegaly, extramedullary hematopoiesis, and body weight loss in *CflipF/F;LysM-Cre* mice. The same group also reported that TNF-α-induced apoptosis is increased in eosinophils of *CflipF/F;LysM-Cre* mice (Gordy et al. 2014). Together, these results suggest that expression of cFLIP is indispensable for cell survival of each hematopoietic lineage by preventing cell death.

6.3 Roles for cFLIP in Intestinal Epithelial Cells

To investigate a role for cFLIP in intestinal epithelial cells (IECs), we and others generated IEC-specific *Cflip*-deficient mice (Table 1) (Piao et al. 2012; Wittkopf et al. 2013). Comparing with IEC-specific *NF-kB essential modulator* (*Nemo*)-, *IkB kinase* (*Ikk*)*b*-, *Fadd*-, or *Caspase-8*-deficient mice (Greten et al. 2004; Gunther et al. 2011; Nenci et al. 2007; Welz et al. 2011), IEC-specific *Cflip*-deficient mice exhibit severe phenotypes (Piao et al. 2012; Wittkopf et al. 2013). IEC-specific *Cflip*-deficient mice die perinatally due to severe colitis along with intestinal bleeding, which is reminiscent of IEC-specific *Tgfb-activated kinase* 1 (*Tak1*)-deficient mice (Kajino-Sakamoto et al. 2008). IECs of tamoxifen-inducible *Cflip*-deficient mice succumb due to severe colitis soon after tamoxifen administration (Weinlich et al. 2013; Wittkopf et al. 2013). Histological analyses of the intestine of IEC-specific *Cflip*-deficient mice reveal that IECs die by apoptosis and necroptosis (Piao et al. 2012), whereas tamoxifen-inducible *Cflip*-deficient IECs mostly die by apoptosis (Weinlich et al. 2013; Wittkopf et al. 2013). Interestingly, apoptosis of IECs is already detected in IEC-specific *Cflip*-deficient mice at E18.5, suggesting that signal(s) other than commensal bacteria might be involved in apoptosis of IECs (Piao et al. 2012). Together, cFLIP plays an essential role in the survival of IECs during perinatal and postnatal stages by preventing apoptosis and necroptosis. Development of severe colitis in IEC-specific *Cflip*-deficient mice and tamoxifen-inducible *Cflip*-deficient mice is rescued by deletion of *Tnfr1* gene and administration of neutralizing antibody against TNF-α, respectively (Piao et al. 2012; Weinlich et al. 2013; Wittkopf et al. 2013). However, some, but not all IEC-specific *Cflip*-deficient mice under *Tnfr1*-deficient background still spontaneously develop colitis after birth (unpublished results), it would be interesting to investigate which signal(s) promote colitis in these mice.

6.4 Roles for cFLIP in Epidermis

Consistent with severe phenotype of IECs-specific *Cflip*-deficient mice, germline deletion of *Cflip* in epidermis results in embryonic lethality (Table 1) (Panayotova-Dimitrova et al. 2013). Moreover, tamoxifen-inducible epidermis-specific *Cflip*-deficient mice develop severe dermatitis due to apoptosis of keratinocytes, and the development of dermatitis is ameliorated by administration of neutralizing antibody against TNF-α (Panayotova-Dimitrova et al. 2013; Weinlich et al. 2013). These results further substantiate that TNF-α-dependent signal plays a dominant role in the execution of cell death in *Cflip*-deficient IECs and keratinocytes. Moreover, consistent with the phenotype of epidermis-specific *Nemo-*, *Fadd-*, and *Tak1*-deficient mice (Bonnet et al. 2011; Omori et al. 2006; Pasparakis et al. 2002), numbers of proliferating keratinocytes are increased in epidermis-specific *Cflip*-deficient mice. These results suggest that dying keratinocytes might release factor(s), resulting in proliferation of nearby keratinocytes. Consistently, a recent study by Kumari et al. reported that oxidative stress-dependent ERK activation induces expression of IL-24, resulting in the development of psoriasis-like dermatitis in epidermis-specific *Ikkb*-deficient mice (Kumari et al. 2013). Therefore, it would be interesting to test whether expression of IL-24 is upregulated in the epidermis of *Cflip*-deficient mice and blockade of IL-24 function might ameliorate dermatitis in *Cflip*-deficient mice.

6.5 Roles for cFLIP in Hepatocytes

In contrast to severe phenotype of IEC-specific or epidermis-specific *Cflip*-deficient mice (Panayotova-Dimitrova et al. 2013; Piao et al. 2012; Weinlich et al. 2013; Wittkopf et al. 2013), hepatocyte-specific *Cflip*-deficient mice by crossing *CflipF/F* mice with *Albumin-Cre* mice do not spontaneously develop hepatitis (Table 1) (Piao et al. 2012; Schattenberg et al. 2011). However, anti-Fas antibody- or D-galactosamine plus lipopolysaccharide (LPS)-induced hepatitis is significantly exacerbated in *CflipF/F;Alb-Cre* mice compared to control mice (Piao et al. 2012; Schattenberg et al. 2011), suggesting that $cFLIP_L$ plays a role in preventing hepatocytes from cell death under stressed conditions. These results might be sharp contrast to hepatocyte-specific *Nemo-* and *Tak1*-deficient mice that spontaneously develop hepatitis and hepatocellular carcinoma (Bettermann et al. 2010; Luedde et al. 2007). Notably, murine hepatocytes express only $cFLIP_L$. A plausible interpretation would be that molecule(s) other than $cFLIP_L$, such as Bcl2 or $Bcl\text{-}x_L$, play a dominant role in suppression of cell death of hepatocytes under steady-state conditions. Alternatively, we cannot formally exclude the possibility that incomplete depletion of *cFlip* in hepatocytes might be responsible for a relatively mild phenotype of *CflipF/F;Alb-Cre* mice. To discriminate these two possibilities, we crossed *CflipF/F* mice with another hepatocyte-specific *Cre*-transgenic-line, *Alpha-fetoprotein (Alfp)-Cre* mice (Kellendonk et al. 2000).

Surprisingly, all *CflipF/F;Alfp-Cre* mice die soon after birth due to massive apoptosis and necroptosis of hepatocytes (Piao et al. 2012). Crossing of *CflipF/F;Alfp-Cre* mice with *Tnfr1-/-* mice does not rescue perinatal lethality, suggesting that death ligand(s) other than TNF-α such as FasL or TRAIL might also contribute to cell death of *Cflip*-deficient hepatocytes (Piao et al. 2012). To investigate a role for cFLIP in adult hepatocytes, we generated polyI:C-inducible *Cflip*-deficient mice by crossing *CflipF/F* mice with *Mx1-Cre* mice. Along with the depletion of $cFLIP_L$ in hepatocytes at protein levels, *CflipF/F;Mx1-Cre* mice, but not control mice, develop fulminant hepatitis and succumb within 3 days after polyI:C injection (Piao et al. 2012). Interestingly, polyI:C-induced hepatitis is blocked by the administration of mixtures of three neutralizing antibodies against anti-TNF-α, TRAIL, and FasL, but not either TNF-α, FasL, or TRAIL alone or in pairwise combinations (Piao et al. 2012), suggesting that cFLIP plays an essential role in the protection of adult hepatocytes from TNF-α-, TRAIL-, and FasL-induced apoptosis.

7 Transgenic Expression of $CFLIP_L$ and *CFLIPs* in T Cells Exhibits Relatively Mild Phenotypes

Transgenic mice overexpressing $CFLIP_L$ or *CFLIPs* specifically in T cells have been generated. T cells from $CFLIP_L$ Tg mice are resistant to anti-Fas Ab-induced apoptosis and undergo Th2 differentiation (Qiao et al. 2010; Tseveleki et al. 2007; Wu et al. 2004). Consistently, Th2-driven allergic reactions such as ovalbumin (OVA)-induced asthma and contact hypersensitivity response are markedly enhanced, but resistant to experimental autoimmune encephalomyelitis in *CFLIPs* Tg mice. Moreover, Qiao et al. reported that $CFLIP_L$ Tg mice spontaneously develop SLE-like autoimmune disease on Balb/c, but not C57BL/6 background (Qiao et al. 2010). Compared to $CFLIP_L$ Tg mice, *CFLIPs* Tg mice exhibit relatively mild phenotype, in which T cell proliferation is impaired but resistant to anti-Fas Ab-induced apoptosis (Hinshaw-Makepeace et al. 2008; Oehme et al. 2005). Together, these results suggest that altered susceptibility of T cells to death receptor-induced apoptosis might promote differentiation of naïve CD4+ T cells into Th2 cells, although the detailed molecular mechanisms are not fully investigated.

Furthermore, testicular germ cell-specific expression of $CFLIP_L$ results in testis atrophy and sperm motility, further substantiating that $CFLIP_L$ has dual functions including suppression and promotion of apoptosis in a cell-type-dependent or context-dependent manner (Antonangeli et al. 2010).

8 Roles for cFLIP in the Development of Human Diseases

Taken that overexpression of cFLIP renders cells resistant to death receptor-induced apoptosis, one might surmise that tumor cells overexpressing cFLIP might be resistant to chemotherapy-induced apoptosis. Consistently, the expression of cFLIP is frequently elevated in many tumors and correlated with tumor progression (Kim et al. 2008; Micheau 2003; Valente et al. 2006). Conversely, knockdown of *CFLIP* in tumor cells by siRNA increases susceptibility to death receptor-induced apoptosis. Moreover, upregulation of cFLIP in peripheral lymphocytes is observed in relapsing multiple sclerosis patients (Gomes et al. 2003). Collectively, manipulation of expression of cFLIP might be a strategy to treat cancer and other diseases that may be caused by alteration of susceptibility to cell death.

9 Is Oxidative Stress Involved in Necroptosis?

9.1 Cellular Sources of ROS Responsible for Necroptosis

We finally discuss the contribution of oxidative stress to necroptosis. Oxidative stress has been shown to be involved in execution of both apoptosis and non-apoptotic cell death; however, the molecular mechanisms are not fully understood (Nakano et al. 2006; Papa et al. 2006). In 1998, Vandenabeele's laboratory reported that TNF-α induces necrosis in murine fibrosarcoma L929 cells, which is suppressed in the presence of an antioxidant, butylated hydroxyanisole (BHA) (Vercammen et al. 1998). We and others have reported that BHA suppresses TNF-α-induced necrosis in NF-κB-deficient fibroblasts (Kamata et al. 2005; Lin et al. 2004; Sakon et al. 2003). We also found that Nec-1 suppresses TNF-α-induced necrosis in these cells, suggesting that NF-κB-deficient cells die, at least in part, by necroptosis upon TNF-α stimulation (Shindo et al. 2013). The fact that oxidative stress is tightly associated with necroptosis raises the following two crucial questions. The first one is how ROS are generated during necroptosis in cells, and the second one is whether ROS are indispensable for the execution of necroptosis.

Regarding the first question, several groups previously reported that TNFR1 is coupled with NADPH oxidase (Nox) on the plasma membrane and Noxs might be source of ROS production (Kim et al. 2007; Yazdanpanah et al. 2009). Indeed, upon various stimuli, Noxs produce superoxide that plays an important role in killing of phagocytosed bacteria and/or transmitting growth signals (Sumimoto 2008). However, at least in our hands, diphenyleneiodonium (DPI), an inhibitor of Noxs did not suppress TNF-α-induced ROS accumulation or cell death in NF-κB-deficient cells (unpublished results), suggesting that Nox-induced ROS do not play a major role in execution of necroptosis. On the other hand, Han's group reported that RIPK3 directly interacts with several enzymes that are involved in glucose metabolism (Zhang et al. 2009a). Facilitation of glycolysis along with the tricarboxylic acid (TCA) cycle increases electron flow involved in oxidative phosphorylation, thus resulting in overproduction of superoxide as a by-product in mitochondria.

9.2 Contribution of ROS and Mitochondria to Necroptosis

To test the contribution of mitochondria to necroptosis, Tait et al. generated cells stably overexpressing Parkin that eliminates damaged mitochondria by mitophagy (Narendra et al. 2008). Then, they treated the cells with an uncoupler, carbonyl cyanide m-chlorophenylhydrazone (CCCP) to reduce membrane potential of mitochondria, resulting in elimination of mitochondria by Parkin (Narendra et al. 2008; Tait et al. 2013). Intriguingly, TNF-α still induces necroptosis in cells lacking mitochondria, suggesting that mitochondria are dispensable for execution of necroptosis, at least under the authors' experimental conditions. Moreover, mixed lineage kinase domain-like (MLKL), an effector molecule downstream of RIPK3 for necroptosis, became oligomerized and is recruited to the plasma membrane through binding to phosphatidylinositol lipids and cardiolipin upon necroptosis induction (Dondelinger et al. 2014; Hildebrand et al. 2014; Wang et al. 2014). Recruited MLKL subsequently formed cytolytic pores on the plasma membrane, resulting in necroptosis. Taken that the recruitment of oligomerized MLKL to the plasma membrane is sufficient for necroptosis, ROS do not appear to play a major role in the execution of necroptosis. Collectively, these results suggest that mitochondria, and possibly mitochondria-derived ROS, are not essential for the execution of necroptosis. Nevertheless, we and others have reported that BHA almost completely suppresses TNF-α-induced necroptosis in various cells (Lin et al. 2004; Sakon et al. 2003; Vercammen et al. 1998). One might surmise that the effect of BHA on necroptosis is an "off-target" effect. However, this is unlikely, since our structural and functional analysis of BHA showed that antioxidant activity of BHA is tightly correlated with the activity of suppression of necroptosis (Shindo et al. 2013).

Based on the requirement of mitochondria for the execution of apoptosis triggered by death receptors, most cells are classified into two types, designated as type I and type II cells (Scaffidi et al. 1999). In type I cells, activation of caspase-8 is sufficient for activation of caspase-3, resulting in apoptosis without apoptogenic factors released from mitochondria. In sharp contrast, in type II cells, activation of caspase-8 is not sufficient for activation of caspase-3; therefore, tBid-dependent release of cytochrome C from mitochondria is prerequisite for full activation of caspase-3 leading to apoptosis. Consistently, overexpression of Bcl-2 or Bcl-xL inhibits death receptor-induced apoptosis in type II cells, but not type I cells (Scaffidi et al. 1999).

By applying analogy of type I and type II cells to necroptosis, we might divide cells undergoing necroptosis into type N1 (N represents necroptosis) and type N2 cells based on the contribution of mitochondria or mitochondria-derived ROS (Fig. 3). In type N1 cells, such as HT29 cells or NIH3T3 cells, necroptosis is executed without mitochondria and not suppressed by antioxidants (He et al. 2009; Tait et al. 2013). In type N2 cells, such as L929 cells or NF-κB-deficient fibroblasts, mitochondria-derived ROS might be indispensable for necroptosis; therefore, necroptosis is suppressed by antioxidants (Lin et al. 2004; Sakon et al. 2003;

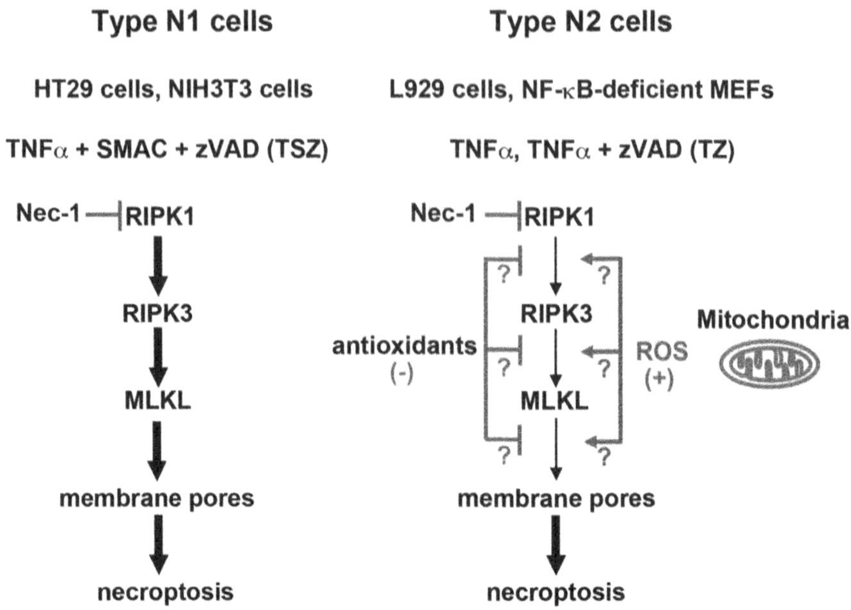

Fig. 3 A model for classification of cells that undergo necroptosis. In type N1 cells, such as HT29 cells or NIH3T3 cells, TSZ stimulation induces necroptosis in ROS-independent manner. On the other hand, in type N2 cells such as L929 cells or NF-κB-deficient cells, TNF-α alone or TZ stimulation induces ROS-dependent necroptosis. Thus, antioxidants such as BHA efficiently block necroptosis. However, the detailed mechanisms by which ROS contribute to necroptosis are currently unknown. TZ; TNF-α + zVAD-fmk, TSZ; TNF-α + SMAC + zVAD-fmk

Vercammen et al. 1998). Although the mechanism how ROS promote necroptosis still remains unclear, this model might explain these apparent inconsistent results. Taken that BHA did not suppress phosphorylation of RIPK1 in TNF-α-stimulated NF-κB-deficient fibroblasts (Shindo et al. 2013), ROS might act as amplifiers downstream RIPK1 activation in type N2 cells. Furthermore, oxidative stress is critically involved in another type of non-apoptotic cell death named ferroptosis (Dixon and Stockwell 2014). It would be intrigued to test whether the signaling pathways leading to necroptosis and ferroptosis might have some cross talk each other. Further study will be required to address this issue.

10 Conclusions

From the initial identification of the first member of the mammalian caspase family named interleukin-1β converting enzyme (ICE) (Miura et al. 1993), the signaling pathways leading to apoptosis have been extensively investigated. Therefore, now we have a near complete picture of how apoptosis is executed and finely tuned by large numbers of signaling molecules. Notably, recent progress in the cell death field has provided two breakthroughs. First, non-apoptotic cell death including necroptosis, pyroptosis, and ferroptosis has been identified, and their biological significances are being established in vitro and in vivo (Bergsbaken et al. 2009; Dixon and Stockwell 2014; Kroemer et al. 2009; Pasparakis and Vandenabeele 2015; Tait et al. 2014). However, the signaling pathways leading to non-apoptotic cell death are not yet fully understood. Secondly, accumulating studies have shown that dying cells release factors that promote inflammation, immunity, and tissue repair (Bergmann and Steller 2010; Zitvogel et al. 2010). Cell death signals and factors released from apoptotic and non-apoptotic cells need further investigation for the better understanding of tissue homeostasis and pathological conditions associated with cell death.

Acknowledgments We thank CF. Ware, M. Miura, S. Yamazaki, and H. Imai for helpful comments on the manuscript. We also thank members of Department of Biochemistry, Toho University School of Medicine, for helpful discussion. RS is supported by a Research Fellowship from Japan Society for the Promotion of Science (JSPS), Japan. This work was supported in part by Grants-in-Aid from Scientific Research (B) (24390100) and Challenging Exploratory Research (25670167) from Japan Society for the Promotion of Science (JSPS), Scientific Research on Innovative areas (26110003) from a MEXT (Ministry of Education, Culture, Sports, Science and Technology), Japan, and research grants from NOVARTIS Foundation for the Promotion of Science, the Naito Science Foundation, the Uehara Science Foundation, and the Takeda Science Foundation.

Competing interests The authors declare that they have no competing interests.

References

Adachi M, Suematsu S, Kondo T, Ogasawara J, Tanaka T, Yoshida N, Nagata S (1995) Targeted mutation in the Fas gene causes hyperplasia in peripheral lymphoid organs and liver. Nat Genet 11:294–300

Antonangeli F, Petrungaro S, Coluccia P, Filippini A, Ziparo E, Giampietri C (2010) Testis atrophy and reduced sperm motility in transgenic mice overexpressing c-FLIP$_L$. Fertil Steril 93:1407–1414

Bergmann A, Steller H (2010) Apoptosis, stem cells, and tissue regeneration. Sci Signal 3:re8

Bergsbaken T, Fink SL, Cookson BT (2009) Pyroptosis: host cell death and inflammation. Nat Rev Microbiol 7:99–109

Bettermann K, Vucur M, Haybaeck J, Koppe C, Janssen J, Heymann F, Weber A, Weiskirchen R, Liedtke C, Gassler N, Muller M, de Vos R, Wolf MJ, Boege Y, Seleznik GM, Zeller N, Erny D, Fuchs T, Zoller S, Cairo S, Buendia MA, Prinz M, Akira S, Tacke F, Heikenwalder M, Trautwein C, Luedde T (2010) TAK1 suppresses a NEMO-dependent but NF-κB-independent pathway to liver cancer. Cancer Cell 17:481–496

Bonnet MC, Preukschat D, Welz PS, van Loo G, Ermolaeva MA, Bloch W, Haase I, Pasparakis M (2011) The adaptor protein FADD protects epidermal keratinocytes from necroptosis in vivo and prevents skin inflammation. Immunity 35:572–582

Budd RC, Yeh WC, Tschopp J (2006) cFLIP regulation of lymphocyte activation and development. Nat Rev Immunol 6:196–204

Chang L, Kamata H, Solinas G, Luo JL, Maeda S, Venuprasad K, Liu YC, Karin M (2006) The E3 ubiquitin ligase itch couples JNK activation to TNF-α-induced cell death by inducing c-$FLIP_L$ turnover. Cell 124:601–613

Chau H, Wong V, Chen NJ, Huang HL, Lin WJ, Mirtsos C, Elford AR, Bonnard M, Wakeham A, You-Ten AI, Lemmers B, Salmena L, Pellegrini M, Hakem R, Mak TW, Ohashi P, Yeh WC (2005) Cellular FLICE-inhibitory protein is required for T cell survival and cycling. J Exp Med 202:405–413

Cretney E, Takeda K, Yagita H, Glaccum M, Peschon JJ, Smyth MJ (2002) Increased susceptibility to tumor initiation and metastasis in TNF-related apoptosis-inducing ligand-deficient mice. J Immunol 168:1356–1361

De Smaele E, Zazzeroni F, Papa S, Nguyen DU, Jin R, Jones J, Cong R, Franzoso G (2001) Induction of gadd45beta by NF-κB downregulates pro-apoptotic JNK signalling. Nature 414:308–313

Degterev A, Hitomi J, Germscheid M, Ch'en IL, Korkina O, Teng X, Abbott D, Cuny GD, Yuan C, Wagner G, Hedrick SM, Gerber SA, Lugovskoy A, Yuan J (2008) Identification of RIP1 kinase as a specific cellular target of necrostatins. Nat Chem Biol 4:313–321

Degterev A, Huang Z, Boyce M, Li Y, Jagtap P, Mizushima N, Cuny GD, Mitchison TJ, Moskowitz MA, Yuan J (2005) Chemical inhibitor of nonapoptotic cell death with therapeutic potential for ischemic brain injury. Nat Chem Biol 1:112–119

Dillon CP, Oberst A, Weinlich R, Janke LJ, Kang TB, Ben-Moshe T, Mak TW, Wallach D, Green DR (2012) Survival function of the FADD-CASPASE-8-$cFLIP_L$ complex. Cell Rep 1:401–407

Dixon SJ, Stockwell BR (2014) The role of iron and reactive oxygen species in cell death. Nat Chem Biol 10:9–17

Dondelinger Y, Declercq W, Montessuit S, Roelandt R, Goncalves A, Bruggeman I, Hulpiau P, Weber K, Sehon CA, Marquis RW, Bertin J, Gough PJ, Savvides S, Martinou JC, Bertrand MJ, Vandenabeele P (2014) MLKL compromises plasma membrane integrity by binding to phosphatidylinositol phosphates. Cell Rep 7:971–981

Feoktistova M, Geserick P, Kellert B, Dimitrova DP, Langlais C, Hupe M, Cain K, MacFarlane M, Hacker G, Leverkus M (2011) cIAPs block Ripoptosome formation, a RIP1/caspase-8 containing intracellular cell death complex differentially regulated by cFLIP isoforms. Mol Cell 43:449–463

Gao M, Labuda T, Xia Y, Gallagher E, Fang D, Liu YC, Karin M (2004) Jun turnover is controlled through JNK-dependent phosphorylation of the E3 ligase Itch. Science 306:271–275

Golks A, Brenner D, Fritsch C, Krammer PH, Lavrik IN (2005) c-$FLIP_R$, a new regulator of death receptor-induced apoptosis. J Biol Chem 280:14507–14513

Goltsev YV, Kovalenko AV, Arnold E, Varfolomeev EE, Brodianskii VM, Wallach D (1997) CASH, a novel caspase homologue with death effector domains. J Biol Chem 272:19641–19644

Gomes AC, Jonsson G, Mjornheim S, Olsson T, Hillert J, Grandien A (2003) Upregulation of the apoptosis regulators cFLIP, CD95 and CD95 ligand in peripheral blood mononuclear cells in relapsing-remitting multiple sclerosis. J Neuroimmunol 135:126–134

Gordy C, Liang J, Pua H, He YW (2014) c-FLIP protects eosinophils from TNF-α-mediated cell death in vivo. PLoS ONE 9:e107724

Gordy C, Pua H, Sempowski GD, He YW (2011) Regulation of steady-state neutrophil homeostasis by macrophages. Blood 117:618–629

Grambihler A, Higuchi H, Bronk SF, Gores GJ (2003) $cFLIP_L$ inhibits p38 MAPK activation: an additional anti-apoptotic mechanism in bile acid-mediated apoptosis. J Biol Chem 278:26831–26837

Greten FR, Eckmann L, Greten TF, Park JM, Li ZW, Egan LJ, Kagnoff MF, Karin M (2004) IKKbeta links inflammation and tumorigenesis in a mouse model of colitis-associated cancer. Cell 118:285–296

Gunther C, Martini E, Wittkopf N, Amann K, Weigmann B, Neumann H, Waldner MJ, Hedrick SM, Tenzer S, Neurath MF, Becker C (2011) Caspase-8 regulates TNF-α-induced epithelial necroptosis and terminal ileitis. Nature 477:335–339

Han DK, Chaudhary PM, Wright ME, Friedman C, Trask BJ, Riedel RT, Baskin DG, Schwartz SM, Hood L (1997) MRIT, a novel death-effector domain-containing protein, interacts with caspases and BclXL and initiates cell death. Proc Natl Acad Sci U S A 94:11333–11338

He S, Wang L, Miao L, Wang T, Du F, Zhao L, Wang X (2009) Receptor interacting protein kinase-3 determines cellular necrotic response to TNF-α. Cell 137:1100–1111

Hildebrand JM, Tanzer MC, Lucet IS, Young SN, Spall SK, Sharma P, Pierotti C, Garnier JM, Dobson RC, Webb AI, Tripaydonis A, Babon JJ, Mulcair MD, Scanlon MJ, Alexander WS, Wilks AF, Czabotar PE, Lessene G, Murphy JM, Silke J (2014) Activation of the pseudokinase MLKL unleashes the four-helix bundle domain to induce membrane localization and necroptotic cell death. Proc Natl Acad Sci U S A 111:15072–15077

Hinshaw-Makepeace J, Huston G, Fortner KA, Russell JQ, Holoch D, Swain S, Budd RC (2008) c-FLIP(S) reduces activation of caspase and NF-κB pathways and decreases T cell survival. Eur J Immunol 38:54–63

Hu S, Vincenz C, Ni J, Gentz R, Dixit VM (1997) I-FLICE, a novel inhibitor of tumor necrosis factor receptor-1- and CD-95-induced apoptosis. J Biol Chem 272:17255–17257

Imamura R, Konaka K, Matsumoto N, Hasegawa M, Fukui M, Mukaida N, Kinoshita T, Suda T (2004) Fas ligand induces cell-autonomous NF-κB activation and interleukin-8 production by a mechanism distinct from that of tumor necrosis factor-α. J Biol Chem 279:46415–46423

Inohara N, Koseki T, Hu Y, Chen S, Nunez G (1997) CLARP, a death effector domain-containing protein interacts with caspase-8 and regulates apoptosis. Proc Natl Acad Sci U S A 94:10717–10722

Irmler M, Thome M, Hahne M, Schneider P, Hofmann K, Steiner V, Bodmer JL, Schroter M, Burns K, Mattmann C, Rimoldi D, French LE, Tschopp J (1997) Inhibition of death receptor signals by cellular FLIP. Nature 388:190–195

Kaiser WJ, Upton JW, Long AB, Livingston-Rosanoff D, Daley-Bauer LP, Hakem R, Caspary T, Mocarski ES (2011) RIP3 mediates the embryonic lethality of caspase-8-deficient mice. Nature 471:368–372

Kajino-Sakamoto R, Inagaki M, Lippert E, Akira S, Robine S, Matsumoto K, Jobin C, Ninomiya-Tsuji J (2008) Enterocyte-derived TAK1 signaling prevents epithelium apoptosis and the development of ileitis and colitis. J Immunol 181:1143–1152

Kamata H, Honda S, Maeda S, Chang L, Hirata H, Karin M (2005) Reactive oxygen species promote TNF-α-induced death and sustained JNK activation by inhibiting MAP kinase phosphatases. Cell 120:649–661

Kataoka T, Budd RC, Holler N, Thome M, Martinon F, Irmler M, Burns K, Hahne M, Kennedy N, Kovacsovics M, Tschopp J (2000) The caspase-8 inhibitor FLIP promotes activation of NF-κB and Erk signaling pathways. Curr Biol 10:640–648

Kavuri SM, Geserick P, Berg D, Dimitrova DP, Feoktistova M, Siegmund D, Gollnick H, Neumann M, Wajant H, Leverkus M (2011) Cellular FLICE-inhibitory protein (cFLIP) isoforms block CD95- and TRAIL death receptor-induced gene induction irrespective of processing of caspase-8 or cFLIP in the death-inducing signaling complex. J Biol Chem 286:16631–16646

Kellendonk C, Opherk C, Anlag K, Schutz G, Tronche F (2000) Hepatocyte-specific expression of Cre recombinase. Genesis 26:151–153

Kim S, Lee TJ, Park JW, Kwon TK (2008) Overexpression of cFLIPs inhibits oxaliplatin-mediated apoptosis through enhanced XIAP stability and Akt activation in human renal cancer cells. J Cell Biochem 105:971–979

Kim YS, Morgan MJ, Choksi S, Liu ZG (2007) TNF-induced activation of the Nox1 NADPH oxidase and its role in the induction of necrotic cell death. Mol Cell 26:675–687

Koenig A, Buskiewicz IA, Fortner KA, Russell JQ, Asaoka T, He YW, Hakem R, Eriksson JE, Budd RC (2014) The c-FLIPL cleavage product p43FLIP promotes activation of extracellular signal-regulated kinase (ERK), nuclear factor κB (NF-κB), and caspase-8 and T cell survival. J Biol Chem 289:1183–1191

Kreuz S, Siegmund D, Scheurich P, Wajant H (2001) NF-κB inducers upregulate cFLIP, a cycloheximide-sensitive inhibitor of death receptor signaling. Mol Cell Biol 21:3964–3973

Kroemer G, Galluzzi L, Vandenabeele P, Abrams J, Alnemri ES, Baehrecke EH, Blagosklonny MV, El-Deiry WS, Golstein P, Green DR, Hengartner M, Knight RA, Kumar S, Lipton SA, Malorni W, Nunez G, Peter ME, Tschopp J, Yuan J, Piacentini M, Zhivotovsky B, Melino G, Nomenclature Committee on Cell D (2009) Classification of cell death: recommendations of the Nomenclature Committee on Cell Death 2009. Cell Death Differ 16:3–11

Kumari S, Bonnet MC, Ulvmar MH, Wolk K, Karagianni N, Witte E, Uthoff-Hachenberg C, Renauld JC, Kollias G, Toftgard R, Sabat R, Pasparakis M, Haase I (2013) Tumor necrosis factor receptor signaling in keratinocytes triggers interleukin-24-dependent psoriasis-like skin inflammation in mice. Immunity 39:899–911

Lin Y, Choksi S, Shen HM, Yang QF, Hur GM, Kim YS, Tran JH, Nedospasov SA, Liu ZG (2004) Tumor necrosis factor-induced nonapoptotic cell death requires receptor-interacting protein-mediated cellular reactive oxygen species accumulation. J Biol Chem 279:10822–10828

Luedde T, Beraza N, Kotsikoris V, van Loo G, Nenci A, De Vos R, Roskams T, Trautwein C, Pasparakis M (2007) Deletion of NEMO/IKKgamma in liver parenchymal cells causes steatohepatitis and hepatocellular carcinoma. Cancer Cell 11:119–132

Micheau O (2003) Cellular FLICE-inhibitory protein: an attractive therapeutic target? Expert Opin Ther Targets 7:559–573

Micheau O, Lens S, Gaide O, Alevizopoulos K, Tschopp J (2001) NF-κB signals induce the expression of c-FLIP. Mol Cell Biol 21:5299–5305

Miura M, Zhu H, Rotello R, Hartwieg EA, Yuan J (1993) Induction of apoptosis in fibroblasts by IL-1 beta-converting enzyme, a mammalian homolog of the C. elegans cell death gene ced-3. Cell 75:653–660

Mocarski ES, Kaiser WJ, Livingston-Rosanoff D, Upton JW, Daley-Bauer LP (2014) True grit: programmed necrosis in antiviral host defense, inflammation, and immunogenicity. J Immunol 192:2019–2026

Naito M, Katayama R, Ishioka T, Suga A, Takubo K, Nanjo M, Hashimoto C, Taira M, Takada S, Takada R, Kitagawa M, Matsuzawa S, Reed JC, Tsuruo T (2004) Cellular FLIP inhibits beta-catenin ubiquitylation and enhances Wnt signaling. Mol Cell Biol 24:8418–8427

Nakagiri S, Murakami A, Takada S, Akiyama T, Yonehara S (2005) Viral FLIP enhances Wnt signaling downstream of stabilized beta-catenin, leading to control of cell growth. Mol Cell Biol 25:9249–9258

Nakajima A, Kojima Y, Nakayama M, Yagita H, Okumura K, Nakano H (2008) Downregulation of c-FLIP promotes caspase-dependent JNK activation and reactive oxygen species accumulation in tumor cells. Oncogene 27:76–84

Nakajima A, Komazawa-Sakon S, Takekawa M, Sasazuki T, Yeh WC, Yagita H, Okumura K, Nakano H (2006) An antiapoptotic protein, c-FLIP$_L$, directly binds to MKK7 and inhibits the JNK pathway. EMBO J 25:5549–5559

Nakano H, Nakajima A, Sakon-Komazawa S, Piao JH, Xue X, Okumura K (2006) Reactive oxygen species mediate crosstalk between NF-κB and JNK. Cell Death Differ 13:730–737

Narendra D, Tanaka A, Suen DF, Youle RJ (2008) Parkin is recruited selectively to impaired mitochondria and promotes their autophagy. J Cell Biol 183:795–803

Nenci A, Becker C, Wullaert A, Gareus R, van Loo G, Danese S, Huth M, Nikolaev A, Neufert C, Madison B, Gumucio D, Neurath MF, Pasparakis M (2007) Epithelial NEMO links innate immunity to chronic intestinal inflammation. Nature 446:557–561

O'Donnell MA, Perez-Jimenez E, Oberst A, Ng A, Massoumi R, Xavier R, Green DR, Ting AT (2011) Caspase 8 inhibits programmed necrosis by processing CYLD. Nat Cell Biol 13:1437–1442

Oberst A, Dillon CP, Weinlich R, McCormick LL, Fitzgerald P, Pop C, Hakem R, Salvesen GS, Green DR (2011) Catalytic activity of the caspase-8-FLIP$_L$ complex inhibits RIPK3-dependent necrosis. Nature 471:363–367

Oehme I, Neumann F, Bosser S, Zornig M (2005) Transgenic overexpression of the Caspase-8 inhibitor FLIP(short) leads to impaired T cell proliferation and an increased memory T cell pool after staphylococcal enterotoxin B injection. Eur J Immunol 35:1240–1249

Omori E, Matsumoto K, Sanjo H, Sato S, Akira S, Smart RC, Ninomiya-Tsuji J (2006) TAK1 is a master regulator of epidermal homeostasis involving skin inflammation and apoptosis. J Biol Chem 281:19610–19617

Palacios C, Yerbes R, Lopez-Rivas A (2006) Flavopiridol induces cellular FLICE-inhibitory protein degradation by the proteasome and promotes TRAIL-induced early signaling and apoptosis in breast tumor cells. Cancer Res 66:8858–8869

Panayotova-Dimitrova D, Feoktistova M, Ploesser M, Kellert B, Hupe M, Horn S, Makarov R, Jensen F, Porubsky S, Schmieder A, Zenclussen AC, Marx A, Kerstan A, Geserick P, He YW, Leverkus M (2013) cFLIP regulates skin homeostasis and protects against TNF-induced keratinocyte apoptosis. Cell Rep 5:397–408

Papa S, Bubici C, Zazzeroni F, Pham CG, Kuntzen C, Knabb JR, Dean K, Franzoso G (2006) The NF-κB-mediated control of the JNK cascade in the antagonism of programmed cell death in health and disease. Cell Death Differ 13:712–729

Pasparakis M, Courtois G, Hafner M, Schmidt-Supprian M, Nenci A, Toksoy A, Krampert M, Goebeler M, Gillitzer R, Israel A, Krieg T, Rajewsky K, Haase I (2002) TNF-mediated inflammatory skin disease in mice with epidermis-specific deletion of IKK2. Nature 417:861–866

Pasparakis M, Vandenabeele P (2015) Necroptosis and its role in inflammation. Nature 517:311–320

Perry WL, Hustad CM, Swing DA, O'Sullivan TN, Jenkins NA, Copeland NG (1998) The itchy locus encodes a novel ubiquitin protein ligase that is disrupted in a18H mice. Nat Genet 18:143–146

Pfeffer K, Matsuyama T, Kundig TM, Wakeham A, Kishihara K, Shahinian A, Wiegmann K, Ohashi PS, Kronke M, Mak TW (1993) Mice deficient for the 55 kd tumor necrosis factor receptor are resistant to endotoxic shock, yet succumb to L. monocytogenes infection. Cell 73:457–467

Piao X, Komazawa-Sakon S, Nishina T, Koike M, Piao JH, Ehlken H, Kurihara H, Hara M, Van Rooijen N, Schutz G, Ohmuraya M, Uchiyama Y, Yagita H, Okumura K, He YW, Nakano H (2012) c-FLIP maintains tissue homeostasis by preventing apoptosis and programmed necrosis. Sci Signal 5:ra93

Qiao G, Li Z, Minto AW, Shia J, Yang L, Bao L, Tschopp J, Gao JX, Wang J, Quigg RJ, Zhang J (2010) Altered thymic selection by overexpressing cellular FLICE inhibitory protein in T cells causes lupus-like syndrome in a BALB/c but not C57BL/6 strain. Cell Death Differ 17:522–533

Riedl SJ, Salvesen GS (2007) The apoptosome: signalling platform of cell death. Nat Rev Mol Cell Biol 8:405–413

Safa AR (2013) Roles of c-FLIP in apoptosis, Necroptosis, and Autophagy. J Carcinog Mutagen Suppl 6:003

Sakamaki K, Inoue T, Asano M, Sudo K, Kazama H, Sakagami J, Sakata S, Ozaki M, Nakamura S, Toyokuni S, Osumi N, Iwakura Y, Yonehara S (2002) Ex vivo whole-embryo culture of caspase-8-deficient embryos normalize their aberrant phenotypes in the developing neural tube and heart. Cell Death Differ 9:1196–1206

Sakon S, Xue X, Takekawa M, Sasazuki T, Okazaki T, Kojima Y, Piao JH, Yagita H, Yagita H, Nakano T (2003) NF-κB inhibits TNF-induced accumulation of ROS that mediate prolonged MAPK activation and necrotic cell death. EMBO J 22:3898–3909

Sanchez-Perez T, Ortiz-Ferron G, Lopez-Rivas A (2010) Mitotic arrest and JNK-induced proteasomal degradation of FLIP and Mcl-1 are key events in the sensitization of breast tumor cells to TRAIL by antimicrotubule agents. Cell Death Differ 17:883–894

Scaffidi C, Schmitz I, Zha J, Korsmeyer SJ, Krammer PH, Peter ME (1999) Differential modulation of apoptosis sensitivity in CD95 type I and type II cells. J Biol Chem 274:22532–22538

Schattenberg JM, Zimmermann T, Worns M, Sprinzl MF, Kreft A, Kohl T, Nagel M, Siebler J, Schulze Bergkamen H, He YW, Galle PR, Schuchmann M (2011) Ablation of c-FLIP in hepatocytes enhances death-receptor mediated apoptosis and toxic liver injury in vivo. J Hepatol 55:1272–1280

Shindo R, Kakehashi H, Okumura K, Kumagai Y, Nakano H (2013) Critical contribution of oxidative stress to TNF-α-induced necroptosis downstream of RIPK1 activation. Biochem Biophys Res Commun 436:212–216

Shu HB, Halpin DR, Goeddel DV (1997) Casper is a FADD- and caspase-related inducer of apoptosis. Immunity 6:751–763

Silke J, Strasser A (2013) The FLIP side of life. Sci Signal 6:pe2

Srinivasula SM, Ahmad M, Ottilie S, Bullrich F, Banks S, Wang Y, Fernandes-Alnemri T, Croce CM, Litwack G, Tomaselli KJ, Armstrong RC, Alnemri ES (1997) FLAME-1, a novel FADD-like anti-apoptotic molecule that regulates Fas/TNFR1-induced apoptosis. J Biol Chem 272:18542–18545

Sumimoto H (2008) Structure, regulation and evolution of Nox-family NADPH oxidases that produce reactive oxygen species. FEBS J 275:3249–3277

Tait SW, Ichim G, Green DR (2014) Die another way–non-apoptotic mechanisms of cell death. J Cell Sci 127:2135–2144

Tait SW, Oberst A, Quarato G, Milasta S, Haller M, Wang R, Karvela M, Ichim G, Yatim N, Albert ML, Kidd G, Wakefield R, Frase S, Krautwald S, Linkermann A, Green DR (2013) Widespread mitochondrial depletion via mitophagy does not compromise necroptosis. Cell Rep 5:878–885

Tang G, Minemoto Y, Dibling B, Purcell NH, Li Z, Karin M, Lin A (2001) Inhibition of JNK activation through NF-κB target genes. Nature 414:313–317

Tenev T, Bianchi K, Darding M, Broemer M, Langlais C, Wallberg F, Zachariou A, Lopez J, MacFarlane M, Cain K, Meier P (2011) The Ripoptosome, a signaling platform that assembles in response to genotoxic stress and loss of IAPs. Mol Cell 43:432–448

Thome M, Martinon F, Hofmann K, Rubio V, Steiner V, Schneider P, Mattmann C, Tschopp J (1999) Equine herpesvirus-2 E10 gene product, but not its cellular homologue, activates NF-κB transcription factor and c-Jun N-terminal kinase. J Biol Chem 274:9962–9968

Thome M, Schneider P, Hofmann K, Fickenscher H, Meinl E, Neipel F, Mattmann C, Burns K, Bodmer JL, Schroter M, Scaffidi C, Krammer PH, Peter ME, Tschopp J (1997) Viral FLICE-inhibitory proteins (FLIPs) prevent apoptosis induced by death receptors. Nature 386:517–521

Tseveleki V, Tsagozis P, Koutsoni O, Dotsika E, Probert L (2007) Cellular FLIP long isoform transgenic mice overcome inherent Th2-biased immune responses to efficiently resolve Leishmania major infection. Int Immunol 19:1183–1189

Valente G, Manfroi F, Peracchio C, Nicotra G, Castino R, Nicosia G, Kerim S, Isidoro C (2006) cFLIP expression correlates with tumour progression and patient outcome in non-Hodgkin lymphomas of low grade of malignancy. Br J Haematol 132:560–570

Varfolomeev EE, Schuchmann M, Luria V, Chiannilkulchai N, Beckmann JS, Mett IL, Rebrikov D, Brodianski VM, Kemper OC, Kollet O, Lapidot T, Soffer D, Sobe T, Avraham KB, Goncharov T, Holtmann H, Lonai P, Wallach D (1998) Targeted disruption of the mouse Caspase 8 gene ablates cell death induction by the TNF receptors, Fas/Apo1, and DR3 and is lethal prenatally. Immunity 9:267–276

Venuprasad K, Elly C, Gao M, Salek-Ardakani S, Harada Y, Luo JL, Yang C, Croft M, Inoue K, Karin M, Liu YC (2006) Convergence of Itch-induced ubiquitination with MEKK1-JNK signaling in Th2 tolerance and airway inflammation. J Clin Invest 116:1117–1126

Vercammen D, Beyaert R, Denecker G, Goossens V, Van Loo G, Declercq W, Grooten J, Fiers W, Vandenabeele P (1998) Inhibition of caspases increases the sensitivity of L929 cells to necrosis mediated by tumor necrosis factor. J Exp Med 187:1477–1485

Wang H, Sun L, Su L, Rizo J, Liu L, Wang LF, Wang FS, Wang X (2014) Mixed lineage kinase domain-like protein MLKL causes necrotic membrane disruption upon phosphorylation by RIP3. Mol Cell 54:133–146

Weinlich R, Oberst A, Dillon CP, Janke LJ, Milasta S, Lukens JR, Rodriguez DA, Gurung P, Savage C, Kanneganti TD, Green DR (2013) Protective roles for caspase-8 and cFLIP in adult homeostasis. Cell Rep 5:340–348

Welz PS, Wullaert A, Vlantis K, Kondylis V, Fernandez-Majada V, Ermolaeva M, Kirsch P, Sterner-Kock A, van Loo G, Pasparakis M (2011) FADD prevents RIP3-mediated epithelial cell necrosis and chronic intestinal inflammation. Nature 477:330–334

Wittkopf N, Gunther C, Martini E, He G, Amann K, He YW, Schuchmann M, Neurath MF, Becker C (2013) Cellular FLICE-like inhibitory protein secures intestinal epithelial cell survival and immune homeostasis by regulating caspase-8. Gastroenterology 145:1369–1379

Wu W, Rinaldi L, Fortner KA, Russell JQ, Tschopp J, Irvin C, Budd RC (2004) Cellular FLIP long form-transgenic mice manifest a Th2 cytokine bias and enhanced allergic airway inflammation. J Immunol 172:4724–4732

Yang JK, Wang L, Zheng L, Wan F, Ahmed M, Lenardo MJ, Wu H (2005) Crystal structure of MC159 reveals molecular mechanism of DISC assembly and FLIP inhibition. Mol Cell 20:939–949

Yazdanpanah B, Wiegmann K, Tchikov V, Krut O, Pongratz C, Schramm M, Kleinridders A, Wunderlich T, Kashkar H, Utermohlen O, Bruning JC, Schutze S, Kronke M (2009) Riboflavin kinase couples TNF receptor 1 to NADPH oxidase. Nature 460:1159–1163

Yeh WC, Itie A, Elia AJ, Ng M, Shu HB, Wakeham A, Mirtsos C, Suzuki N, Bonnard M, Goeddel DV, Mak TW (2000) Requirement for Casper (c-FLIP) in regulation of death receptor-induced apoptosis and embryonic development. Immunity 12:633–642

Yu JW, Jeffrey PD, Shi Y (2009) Mechanism of procaspase-8 activation by c-$FLIP_L$. Proc Natl Acad Sci U S A 106:8169–8174

Yuan J (2006) Divergence from a dedicated cellular suicide mechanism: exploring the evolution of cell death. Mol Cell 23:1–12

Zhang DW, Shao J, Lin J, Zhang N, Lu BJ, Lin SC, Dong MQ, Han J (2009a) RIP3, an energy metabolism regulator that switches TNF-induced cell death from apoptosis to necrosis. Science 325:332–336

Zhang H, Rosenberg S, Coffey FJ, He YW, Manser T, Hardy RR, Zhang J (2009b) A role for cFLIP in B cell proliferation and stress MAPK regulation. J Immunol 182:207–215

Zhang H, Zhou X, McQuade T, Li J, Chan FK, Zhang J (2011) Functional complementation between FADD and RIP1 in embryos and lymphocytes. Nature 471:373–376

Zhang J, Cado D, Chen A, Kabra NH, Winoto A (1998) Fas-mediated apoptosis and activation-induced T-cell proliferation are defective in mice lacking FADD/Mort1. Nature 392:296–300

Zhang N, He YW (2005) An essential role for c-FLIP in the efficient development of mature T lymphocytes. J Exp Med 202:395–404

Zitvogel L, Kepp O, Kroemer G (2010) Decoding cell death signals in inflammation and immunity. Cell 140:798–804

Lipid Peroxidation-Dependent Cell Death Regulated by GPx4 and Ferroptosis

Hirotaka Imai, Masaki Matsuoka, Takeshi Kumagai, Taro Sakamoto and Tomoko Koumura

Abstract Glutathione peroxidase 4 (Phospholipid hydroperoxide glutathione peroxidase, PHGPx) can directly reduce phospholipid hydroperoxide. Depletion of GPx4 induces lipid peroxidation-dependent cell death in embryo, testis, brain, liver, heart, and photoreceptor cells of mice. Administration of vitamin E in tissue specific GPx4 KO mice restored tissue damage in testis, liver, and heart. These results indicate that suppression of phospholipid peroxidation is essential for cell survival in normal tissues in mice. Ferroptosis is an iron-dependent non-apoptotic cell death that can elicited by pharmacological inhibiting the cystine/glutamate antiporter, system Xc^- (type I) or directly binding and loss of activity of GPx4 (Type II) in cancer cells with high level RAS-RAF-MEK pathway activity or p53 expression, but not in normal cells. Ferroptosis by Erastin (Type I) and RSL3 (RAS-selective lethal 3, Type II) treatment was suppressed by an iron chelator, vitamin E and Ferrostatin-1, antioxidant compound. GPx4 can regulate ferroptosis by suppression of phospholipid peroxidation in erastin and RSL3-induced ferroptosis. Recent works have identified several regulatory factors of erastin and RSL3-induced ferroptosis. In our established GPx4-deficient MEF cells, depletion of GPx4 induce iron and 15LOX-independent lipid peroxidation at 26 h and caspase-independent cell death at 72 h, whereas erastin and RSL3 treatment resulted in iron-dependent ferroptosis by 12 h. These results indicated the possibility that the mechanism of GPx4-depleted cell death might be different from that of ferroptosis induced by erastin and RSL3.

Abbreviations

GPx4	Glutathione peroxidase
mGPx4	Mitochondrial GPx4
cGPx4	Non-mitochondrial GPx4
nGPx4	Nucleolar GPx4

H. Imai (✉) · M. Matsuoka · T. Kumagai · T. Sakamoto · T. Koumura
Department of Hygienic Chemistry, School of Pharmaceutical Sciences, Kitasato University, 5-9-1 Shirokane, Minato-ku, Tokyo 108-8641, Japan
e-mail: imaih@pharm.kitasato-u.ac.jp

Current Topics in Microbiology and Immunology (2017) 403:143–170
DOI 10.1007/82_2016_508

Published Online: 14 February 2017

tBid	Truncated Bid
CL	Cardiolipin
TG	Transgenic
MEF	Mouse embryonic fibroblast
RSL	Ras-Selective Lethal
Tam	Tamoxifen
GSH	Glutathione
LOX	Lipoxygenase
DFO	Deferoxamine
Fer-1	Ferrostatin-1
LOX	Lipoxygenase
PC	Phosphatidylcholine
PE	Phosphatidylethanolamine
ACSL4	Acyl-CoA synthetase long-chain family member 4
AA	Arachidonic acid

Contents

1 Introduction

Ferroptosis is an iron- and lipid peroxidation-dependent and caspase-independent novel form of regulated cell death (RCD), which was recently named in 2012 by Dr. Brent R. Stockwell (Dixon et al. 2012). Ferroptosis is distinct from other types of cell death such as apoptosis, autophagic cell death, and necroptosis (Dixon et al. 2012; Yang and Stockwell 2008). Ferroptosis inducer, Ras-selective lethal small molecule (RSL) including erastin and RSL-3 were found by high-throughput small molecule-screening that selectively killed the mutant Ras oncogene transformed human foreskin fibroblasts (BJeJR), but not their isogenic primary counterparts and normal cells in a non-apoptotic manner (Yang and Stockwell 2008; Yagoda et al. 2007). Multiple inhibitors of apoptosis, necrosis, and autophagy (e.g., Z-VAD-FMK, Boc-D-FMK, wortmannin, and necrostatin-1) cannot rescue ferroptosis by erastin and RSL3 treatment. In contrast, antioxidants [e.g., vitamin E and butylated hydroxytoluene (BHT)] and iron chelator (deferoxamine mesylate) inhibit RSLs induced cell death. These results indicated that ferroptosis refers to an iron-dependent, non-apoptotic cell death.

Glutathione peroxidase 4 (GPx4, Phospholipid hydroperoxide glutathione peroxidase PHGPx) is a unique intracellular antioxidant enzyme that directly reduces peroxidized phospholipids that have been produced in cell membrane (Imai and Nakagawa 2003). GPx4 knockout mice displayed early embryonic lethal in mice at 7.5 dpc (Imai et al. 2003a) and cell death in several tissues of conditional knockout mice (Imai 2010; Seiler et al. 2008). Before reports about ferroptosis, ablation of GPx4 induces lipid peroxidation-dependent, caspase-independent cell death in embryo and MEF cells (Imai et al. 2009; Seiler et al. 2008). GPx4 recently reported to be a regulator of ferroptosis by RSL3 and erastin, since RSL3 and erastin decreased the activity of GPx4 by direct binding to GPx4 and indirectly loss of glutathione respectively (Yang et al. 2014; Dixon et al. 2012).

On the other hand, mitochondrial GPx4 previously reported to be a suppressor of apoptosis by mitochondrial death pathway, since mitochondrial GPx4 inhibits the release of cytochrome c from mitochondria by reduction of cardiolipin hydroperoxide in apoptosis (Nomura et al. 1999, 2000).

In this review, we summarize recent studies on the lipid peroxidation-dependent cell death such as apoptosis and ferroptosis regulated by organelle-specific GPx4 from lessons of analysis of GPx4 overexpressed cells, knockout cells, and mice.

2 Structure and Expression of Three Types of Organelle-Specific GPx4

The GPxs family consists of various members, including GPx1–8 (Imai 2010; Imai and Nakagawa 2003). GPx1, 2, 3, 4, and 6 are selenoproteins that have selenocysteine at active site in human, except cysteine in GPx6 in mice. GPx5, 7 and 8 have very low glutathione peroxidase activity and thioredoxin-like activity because they have cysteine at active site. GPx3 exists in plasma, and GPx2 is expressed in gastrointestinal tract.

GPx1 and GPx4 are generally expressed in normal tissues. GPx1 can reduce hydrogen peroxide and fatty acid hydroperoxide in cytosol, but not phospholipid hydroperoxide in membrane using glutathione. GPx4 can effectively reduce phospholipid hydroperoxide, fatty acid hydroperoxide, cholesterol hydroperoxide, and thymine hydroperoxide, but ineffectively hydrogen peroxide using glutathione.

Three types of GPx4 are transcribed by different transcriptional start codons and different exons (exon1a and exon1b) from one gene (Imai and Nakagawa 2003; Imai et al. 2006). Only N-terminal sequence including signal sequence for transport to different organelle such as mitochondria, nucleoli, cytosol, and nuclei is different among three types of GPx4, whereas the remaining amino acid sequence on the C-terminal side including the enzymatic active site is exactly the same. Thus, GPx4 contain three types of GPx4, mGPx4 that transported into mitochondria (Arai et al. 1996), nGPx4 that localized in nucleoli (Nakamura et al. 2003) and cGPx4 that distributed in cytosolic and nuclei (Imai et al. 1995). cGPx4 might also strongly associate with membrane in cytosolic sites of organelle (Arai et al. 1996) (Fig. 1).

The expression of cGPx4 mRNA was relatively high in somatic cells while that of nGPx4 mRNA was extremely low. In somatic tissues and cultured cells, the amounts of cGPx4 mRNA were approximately 2.5–12, 600–9000 times higher than those of mGPx4 mRNA and nGPx4 mRNA as determined by TaqMan assay. The expression of mGPx4 mRNA and nGPx4 mRNA was significantly higher only in testis than in other tissues. Expression of mGPx4 mRNA and nGPx4 mRNA is induced significantly in testis during spermatogenesis (Imai et al. 2006).

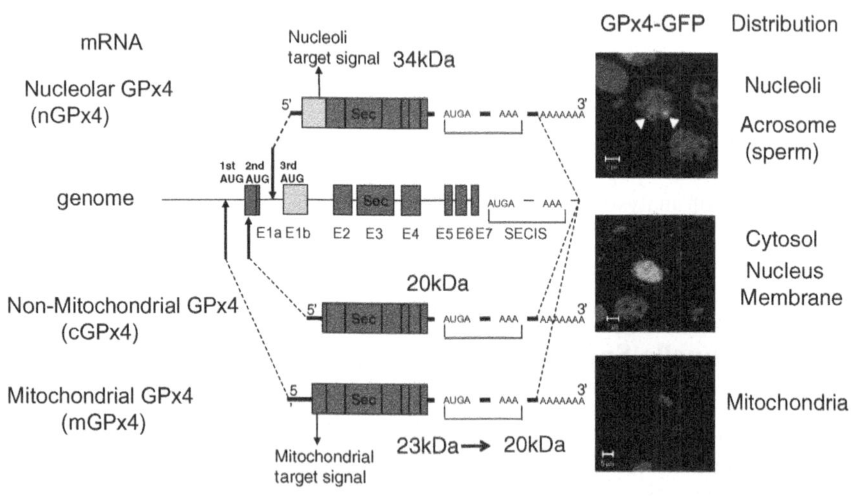

Fig. 1 Structure and distribution of thee types of GPx4. Three types of GPx4, nucleolar, mitochondrial and non-mitochondrial isoforms, are transcribed from one gene by alternative transcription. GPx4 proteins contain nucleolar (34 kDa), non-mitochondrial (20 kDa) and mitochondrial GPx4 (23 kDa) with targeting signal (exon 1b and 1a) at the N-terminal of protein. The parts of C-terminal of three types of GPx4 including selenocysteine (Sec) at the enzymatic active site are the same. GPx4 is one of the selenoproteins including selenocysteine that is encoded by stop codon AUG. Selenocysteine insertion sequence (SECIS) is required for the incorporation of selenocysteine into AUG codon of the GPx4 protein. *Right panels* showed the distribution of each GPx4-GFP fusion protein in rat basophile leukemia (RBL2H3) cells. *Green* GFP fluorescence, *Blue* DAPI staining of nucleus

3 Function of Three Types of Organelle-Specific GPx4 in Cells

Analysis of overexpression of three types of GPx4 in rat basophile leukemia (RBL2H3) cells revealed that three types of GPx4 play organelle-specific independent roles in the modulation of inflammation, signal transduction, and cell death (Imai and Nakagawa 2003; Imai 2010) (Fig. 2).

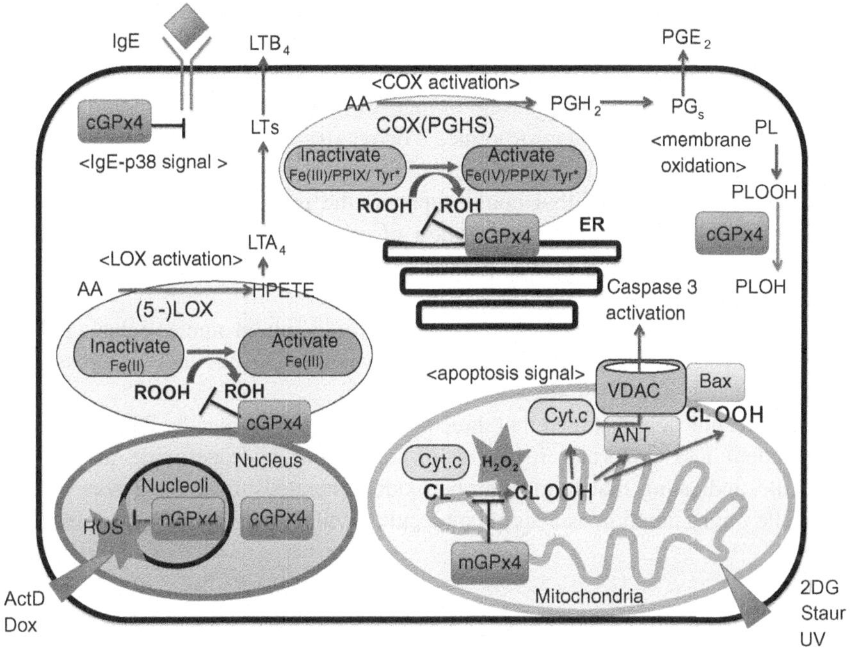

Fig. 2 Functions of three types of GPx4 in rat basophile leukemia cells. Functional analyses of stable transformants overexpressing of three types of GPx4 such as cGPx4 (L9 cells), mGPx4 (M15 cells) and nGPx4 (N63 cells) in rat basophile leukemia cells (RBL2H3 cells) were summarized. GPx4 could directly reduce phospholipid hydroperoxide (PLOOH) to hydroxyl lipid (PLOH) using glutathione (GSH). cGPx4 in the nucleus inhibited the activation of 5-lipoxygenase (5-LOX) by reduction of lipid hydroperoxide (ROOH) to hydroxyl lipid (ROH) as the activator of 5-LOX that is oxidized from inactive Fe (II) to active Fe (III), resulting in the suppression of production of leukotriene B_4 (LTB_4). cGP4 in the endoplasmic reticulum (ER) inhibit the activation of cyclooxygenase (COX), prostaglandin H2 synthase (PGHS) by reduction of lipid hydroperoxide as the activator of COX that is oxidized from inactive heme Fe (III) to active heme Fe (IV), resulting in the suppression of prostaglandin E2 (PGE_2). cGPx4 also suppress the activation of the IgE signaling via p38, resulting in the inhibition of production of platelet-activating factor (PAF). mGPx4 could suppress the release of cytochrome c (cyt.c), the activator of caspase 3 from mitochondria in apoptosis induced by mitochondrial death pathway such as 2-deoxyglucose (2DG), UV and staurosporine (Staur). mGPx4 inhibited the detachment of cytochrome c (cyt.c) from cardiolipin (CL), mitochondrial specific phospholipid and the conformational change of adenine nucleotide transporter (ANT) regulating the opening of permeability transition pore such as voltage-dependent anionic channel (VDAC)-Bax complex by reducing of cardiolipin hydroperoxide (CLOOH). nGPx4 could suppress the oxidative damage of nucleoli induced by doxorubicin (Dox) and actinomycin D (ActD). Three type of organelle-specific GPx4 played an independent and important regulator of local lipid hydroperoxide as a signal molecule. *AA* arachidonic acid, *HPETE* hydroperoxyeicosapentanoicacid, *PL* phospholipid

Overexpression of cGPx4 in the cytosol and nucleus could suppress the production of leukotriene and prostaglandin at the nucleus and endoplasmic reticulum in response to several stimuli, indicating that cGPx4 could suppress the activation of lipoxygenase and cyclooxygenase at the nucleus and endoplasmic reticulum by reducing fatty acid hydroperoxide as activators of lipoxygenase and cyclooxygenase (Imai et al. 1998; Sakamoto et al. 2000).

Overexpression of cGPx4 could suppress the production of platelet-activating factor (PAF) by IgE-antigen stimulation. cGPx4 could suppress the phosphorylation of p38 by lipid hydroperoxide signaling pathway (Sakamoto et al. 2002). However, cGPx4 could not inhibit suppress apoptosis induced by the mitochondrial death pathway (Imai et al. 1996; Arai et al. 1999; Nomura et al. 1999).

On the other hand, overexpression of mGPx4 in the mitochondria could suppress the apoptosis induced by staurosporine, 2-deoxyglucose and UV, whereas mGPx4 could not inhibit the production of leukotriene and prostaglandin (Arai et al. 1999; Nomura et al. 1999). mGPx4 could suppress the release of cytochrome c from mitochondria by inhibition of generation of mitochondria specific phospholipid, cardiolipin hydroperoxide during apoptosis induced by mitochondria death pathway (Nomura et al. 2000).

Overexpression of nGPx4 in the nucleoli could inhibit nucleoli-damaged cell death induced by doxorubicin and actinomycin D, but not staurosporine, 2-deoxyglucose, and UV (Nakamura et al. 2003).

These results demonstrated that three types of organelle-specific GPx4 played an independent and important regulator of local lipid hydroperoxide as a signal molecule, indicating that lipid hydroperoxide generated in the local area of each organelle might function as a signal molecule in inflammation and cell death (Imai 2010).

4 Mitochondrial GPx4 Inhibit the Release of cyt.c from Mitochondria by Cardiolipin Hydroperoxide in Apoptosis by Mitochondrial Death Pathway

Our mGPx4 transformant studies revealed that mGPx4 inhibits apoptosis induced by 2-deoxyglucose, staurosporine, actinomycin D, and UV, but not A23187 and anti-Fas antibody stimulation and mGPx4 could suppress the release of cytochrome c from mitochondria, resulting in inhibition of caspase-3 activation and PS externalization (Arai et al. 1999; Nomura et al. 1999) (Fig. 2).

Cardiolipin, that is located primarily in the mitochondrial inner membrane, has a unique structure containing three glycerol moieties, two phosphate residues, and four fatty acyl chains in the same molecule. Cardiolipin (CL) is critical for maintenance of cristae structure as well as stabilizing mitochondrial electron transport complexes, carrier proteins, and phosphokinases (Imai and Nakagawa 2003; Maguire et al. 2017).

Oxidative stress induced peroxidation of CL because CL is rich in polyunsaturated fatty acids, especially linoleic acid. We found that during apoptosis by mitochondrial pathway, oxidation of CL in the mitochondria occurs early events before the release of cytochrome c from mitochondria. Cytochrome c strongly associates with CL, whereas oxidation of CL is a required step for the dissociation of cytochrome c from CL in the inner membrane, since oxidized CL exhibits a reduced binding affinity for cytochrome c over CL. mGPx4 could suppress the peroxidation of cardiolipin in mitochondria and inhibit the dissociation of cytochrome c from mitochondrial inner membrane in apoptosis (Nomura et al. 2000). And mGPx4 could inhibit the change of conformation by loss of activity of adenine nucleotide translocator (ANT) that could regulate the opening of permeability transition pore by cardiolipin hydroperoxide (Imai et al. 2003b). We proposed "cardiolipin hydroperoxide cascade" for the release of cytochrome c from mitochondria in apoptosis in 2003 (Imai and Nakagawa 2003).

Kagan's group found that pro-apoptotic stimuli induced H_2O_2 dependent peroxidation of CL by cytochrome c (Kagan et al. 2005). Recent works indicated that oxidation of CL results in migration of CL from inner membrane to the outer mitochondrial membrane (Maguire et al. 2017; Li et al. 2015). Oxidized CL recruits and interacts with Bax to initiate formation of mitochondrial transition pore (Korytowski et al. 2011).

Recently our data showed that mGPx4 could suppress the release of cytochrome c from mitochondria by different mechanisms of Bax- and tBid-induced apoptosis (Imai, unpublished). These data demonstrate that mGPx4 is a regulator of apoptosis by inhibition of CL peroxidation-dependent release of cytochrome c from mitochondria.

5 Suppression of Peroxidation of Phospholipid by Non-mitochondrial GPx4 and Vitamin E Is Essential for the Survival of Cells and Mice

Ablation of all GPx4 gene in mice induced early embryonic lethal at 7.5 dpc (Imai et al. 2003a). GPx4-null embryo could not develop into Inner Mass Cell (ICM), whereas vitamin E could rescue formation of ICM in GPx4-null embryo. Transfection of cDNA for cGPx4 also could recover the formation of ICM in GPx4-null embryo. These results indicated that suppression of generation of lipid hydroperoxide by GPx4 and vitamin E is required for embryo development.

We also succeeded that transgenic complementary rescue method using an all GPx4-loxP transgene rescued embryonic lethality in endogenous all GPx4 KO mice (Imai 2010; Imai et al. 2009). mGPx4 starts codon mutation transgene and nGPx4 start codon mutation transgene could rescue the embryonic lethality of all GPx4 KO mice, whereas cGPx start codon mutation transgene could not rescue the embryonic lethality. Double GPx4 mutation transgene containing double mutation of start codons for mGPx4 and nGPx4 could rescue the lethal phenotype, indicating that cGPx4 is essential for embryo development and normal growth in mice (Imai 2011).

In fact, mGPx4 KO mice and nGPx4 KO mice display normal development except sperm maturation (Schneider et al. 2009; Conrad et al. 2005). These results demonstrated that anti-apoptotic function of mGPx4 is not required for embryonic normal development and programed cell death such as apoptosis in mice and cGPx4 is important role for embryo development.

mGPx4 KO mice showed male infertility by structural damage of mitochondria of spermatozoa, but showed normal production of the number of spermatozoa in testis (Schneider et al. 2009; Imai 2011). On the other hand, spermatocyte—specific all GPx4 KO mice showed severe defect of spermatogenic cells in testis, the significant low level of the number of spermatozoa, a hairpin-like flagella and abnormal structure of mitochondria of spermatozoa, resulting in male infertility (Imai et al. 2009; Fujii and Imai 2014). Administration of vitamin E with spermatocyte-specific all GPx4 KO mice could rescue the defect of spermatogenesis in seminiferous tables in mice, leading to the recover of production of the number of spermatozoa.

Docosahexaenoic acid (DHA) is a long-chain omega-3 polyunsaturated fatty acid that is a critical component of lid structure. DHA plays important roles throughout the body and is essential for maintaining the structure and function of the brain and eye. In the rod cells of retinal photoreceptors for example, DHA within membrane facilitates the conformational change triggered by a light signal. However, DHA is easy for oxidation, since it is polyunsaturated fatty acid. Although photoreceptor cells normally developed and differentiated into rod and cone cells by P12 in photoreceptor cell specific GPx4 KO mice, they rapidly underwent drastic degeneration and completely disappeared by P21. Photoreceptor cell death induced by loss of GPx4 was TUNEL positive, lipid oxidation dependent, and caspase-independent cell death. GPx4 is a critical antioxidant enzyme for the maturation and survival of photoreceptor cells (Ueta et al. 2012). Using in vivo wound repair model by application of *n*-heptanol on the corona or laser-induced choroidal neovascularization mice model, we clarified that GPx4 plays important role for oxidative homeostasis and cell survival in the corneal epithelial cells and the retinal pigment epithelium (RPE)/choroid tissue (Sakai et al. 2016a; Roggia et al. 2014).

Tamoxifen inducible all GP×4 KO mice induced acute renal failure with ferroptotic cell death (Angeli et al. 2014).

Heart specific all GPx4 KO mice showed embryonic death by cardio sudden death. However heart specific all GPx4 KO mice could normally grow by administration with vitamin E diet. Heart specific all GPx4 KO mice rescued by vitamin E diet show cardio sudden death by exchange to the normal diet (Imai unpublished).

Liver specific all GPx4 KO mice also showed neonatal lethal, however liver specific all GPx4 KO mice normally grow by administration of vitamin E. And change of vitamin E diets to vitamin E deficient diet induces sudden death of Liver specific all GPx4 KO mice (Imai 2011; Carlson et al. 2016).

Endothelium-specific deletion of GPx4 had no obvious impact on normal vascular homeostasis in mice maintained on a normal diet. However when mice were fed a vitamin E depleted diet for 6 weeks before endothelial deletion of GPx4 was induced by tamoxifen, 80% of endothelium GPx4 knockout mice died with detachment of endothelial cells from the basement membrane (Wortmann et al. 2013).

These results demonstrated that suppression of phospholipid peroxidation by cGPx4 and vitamin E is essential for survival in certain cells and mice and imbalance of suppression of phospholipid peroxidation might cause the several diseases (Fig. 3).

Cell death by genetic depletion of all GPx4 in each tissue of mice is dependent of loss of cGPx4, since mGPx4 and nGPx4 double KO mice normally survive.

In human, a defect of expression of GPx4 in spermatozoa was found in human infertile male with oligoasthenozoospermia that have the low number and low motility of spermatozoa (Imai et al. 2001). This phenotype in human GPx4 deficient

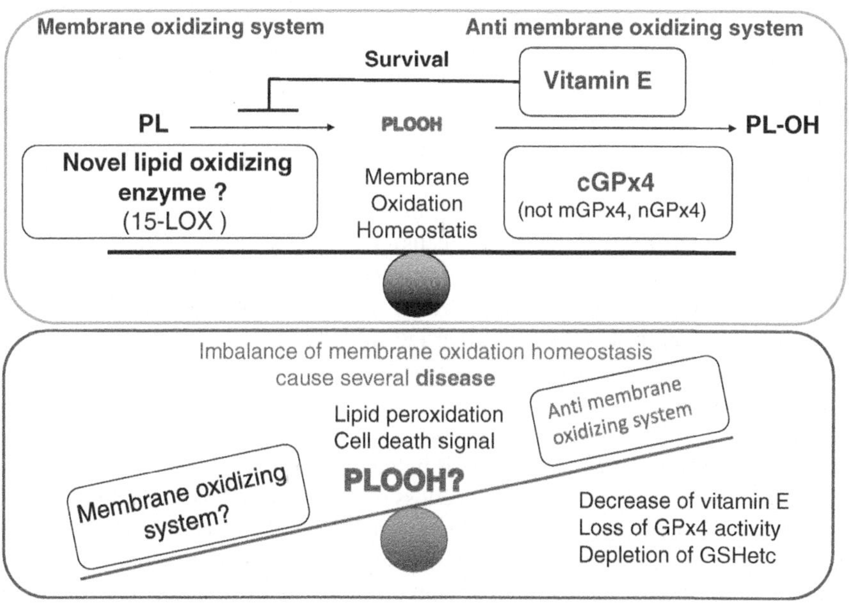

Fig. 3 Imbalance of membrane oxidation homeostasis cause several diseases in cGPx4 knockout mice and cGPx4 depleted cells. GPx4 knockout mice is early embryonic lethal at 7.5 dpc. GPx4 KO embryo could not form Inner cell Mass (ICM), but addition of vitamin E could rescue it. cGPx4 is required for the survival for the normal embryogenesis, since mGPx4 and nGPx4 KO mice normally grow. Testis, Liver specific GPx4 KO mice display the cell death of spermatogenic cells and hepatocyte, but administration of vitamin E diet could rescue the cell death in tissues of mice. 15-Lipoxygenase (15-LOX) directly could oxidize phospholipid. However, GPx4 depletion in 15-LOX KO mice induce acute renal injury and GPx4 depletion in 15-LOX null-MEF cells induce the lipid peroxidation-dependent cell death, indicating that 15-LOX is not essential for lipid peroxidation-dependent cell death in mice. 15-LOX might be one of the candidate for lipid peroxidation-dependent cell death in 15-LOX expressing cells. Thus, imbalance of membrane oxidation homeostasis (membrane oxidizing system vs. anti membrane oxidizing system) cause the oxidative stress related diseases in animal and human. *PLOOH* phospholipid hydroperoxide

oligoasthenozoospermia were consistent with that of spermatocyte-specific all GPx4 KO mice, indicating that low production of spermatozoa in human infertile patient is due to the defect of spermatogenesis by the deficiency of cGPx4 in testis (Imai et al. 2009).

Sedaghahatian-type spondylometaphyseal dysplasia (SSMD) is a neonatal lethal from spondylometaphyseal dysplasia characterized by severe metaphysical chondrodysplasia with limb shortening, platyspondyly, cardiac conduction, defects, and central nerves system abnormalities. By whole exam sequencing of a child affected with SSMD and her unaffected parents, two rare variants of GPx4 that the mutation results in a frameshift and premature truncation of GPx4 were identified (Smith et al. 2014).

These results indicate that truncating mutation in GPx4 in two families affected with SSMD supports the pathogenic role of mutated GPx4 in this vary rare disease.

GPx4 is one of the selenoproteins in human. Selenocysteine insertion sequence-binding protein 2 (SBP2) is essential for the biosynthesis of selenoproteins including GPx4. Subjects with mutations in the SBP2 gene have decreased levels of many selenoproteins, resulting in a several phenotype with high lipid peroxidation in blood since they have low levels of antioxidant activity such as GPx1 and GPx4. Treatment of the vitamin E for 2 years to the subjects with SBP2 mutation reduced lipid peroxidation product levels to the control subjects, indicating that vitamin E treatment effectively inhibits the generation of lipid peroxidation products in human (Saito et al. 2015).

To clarify the mechanism of cell death by depletion of GPx4, we established GPx4-LoP TG/KO MEF cells (TK cells) from GPx4-loxP TG/KO mice and tamoxifen inducible GPx4 depleted TK cells (ETK cells) by infection of retrovirus of estrogen receptor binding Cre (CreERT2) (Imai et al. 2009; Imai 2010). Conrad' group already established the tamoxifen inducible GPx4 depleted MEF cells (Pfa1 cells) from GPx4 flox/flox mice (Seiler et al. 2008).

Retrovirus-mediated depletion of GPx4-LoxP TG transgene in EK cells resulted in the loss of all GPx4 protein 2 days and cell death 4 days after infection of Cre-expressing retrovirus. In tamoxifen inducible all GPx4 depleted TK cells and Pfa1 cells, addition of tamoxifen induce loss of all GPx4 24 h and cell death 3 days after tamoxifen treatment. As characteristics of GPx4-deficient cell death in MEF cells and mice, the time required for lethality is very long. Addition of Trolox, vitamin E derivative, or vitamin E could rescue GPx4 depleted cell death by retrovirus transfection in TK cells and addition of tamoxifen in ETK cells and Pfa1cells. Vitamin E rescued GPx4 null-MEF cells could normally grow.

Retrovirus-infection of mGPx4, cGPx4, nGPx, cGPx4 (cys) and other antioxidants such as GPx1, SOD1, and SOD2 demonstrated that cGPx4 most effectively could rescue the all GPx4 depleted cell death, but not cGPx4 (cys) that mutated a enzymatic active site selenocysteine to cysteine, mGPx4, nGPx4, GPx1, SOD1, and SOD2 (Imai et al. 2009; Imai 2010).

These results demonstrated that the suppression of phospholipid peroxidation by cGPx4 and vitamin E is required for the growth and survival in MEF cells as the same as in mice. GPx4 depletion by genetic system induces lipid peroxidation-dependent, and caspase 3 independent novel cell death, since broad caspase-inhibitor, Z-Bad-FMK could not suppress GPx4 depleted cell death.

Lipid peroxidation by loss of GPx4 was generated in the cytosol sites of organelle, but not in mitochondria, as the signal for novel non-apoptotic cell death (Imai 2010, 2011).

6 Ferroptosis by Oncogenic Mutated Ras-Selective Lethal Compounds

The RAS family of small GTPase (HRAS, NRAS, and KRAS) is commonly mutated in cancer and several groups have searched for small molecules that are selectively lethal to cells expressing oncogenic mutated RAS proteins. Stockewell's group isolated two classes of novel oncogenic RAS-Selective Lethal (RSL) small molecules named eradicator of RAS and ST (erastin) (Yagoda et al. 2007) and RAS-selective Lethal 3 ((1S, 3R)-RSL3) (Yang and Stockwell 2008). Both compounds killed engineered human tumor cells expressing oncogenic $HRAS^{V12}$ at lower concentrations than isogenic cells expressing wild-type HRAS. Erastin and RSL3 treatment could induce cell death very quickly for 8–12 h in RAS-mutated cancer cells. Erastin and RSL3 treatments do not trigger morphological changes or biochemical processes consistent with apoptosis such as chromatin condensation, DNA ladder formation, and caspase 3 activation. Erastin and RSL3 induced cell death is not inhibited by caspase-inhibitor (Z-VAD-FMK), by a necroptosis inhibitor (necrostatin-1) and by an inhibitor of autophagy (chloroquine, 3-methyladenine). Neither mitochondrial ROS production nor Ca^{2+} influx is necessary for cell death. Erastin treatment resulted in a unique "dysmorphic" mitochondrial phenotype observable by transmission electron microscopy. Erastin and RSL3 induced cell death is effectively suppressed by the iron chelators, DFO (deferoxamine), and ciclopirox (CPX) as well as by the lipophilic antioxidants, Trolox (a soluble vitamin E analog), butylated hydroxyanisole (BHA), butylated hydroxytoluene (BHT), liproxstatin-1, and ferrostatin-1 (Fer-1). These results indicated that erastin- or RSL3-induced cell death is the iron-dependent oxidative novel non-apoptotic cell death, named "ferroptosis" (Dixon et al. 2012). Ferrostatin-1 (Fer-1) antioxidants was isolated as a ferroptosis specific inhibitor by Stockwell'group. Ferrostatin-1 could suppress the ferroptosis by erastin and RSL3, but not apoptosis and necroptosis (Dixon et al. 2012).

The elucidation of target molecule for Erastin and RSL3 provides that ferroptosis inducer can be divided into two classes of small molecule substrates. Class 1 ferroptosis inducers include erastin, sulfasalazine (SAS), artemisinin and its derivatives, which can inhibit system X_c^- that transport cystine into cells (Dixon et al. 2012). Class 2 ferroptosis inducers include Ras-selective lethal 3 compound (RSL3), ML162 (DPI17), DPI20, DP112, DP113, etc., which can directly inhibit glutathione peroxidase 4 (GPx4) activity and ultimately lead to an accumulation of lipid peroxides without the decrease of GSH (Yang et al. 2014) (Fig. 4).

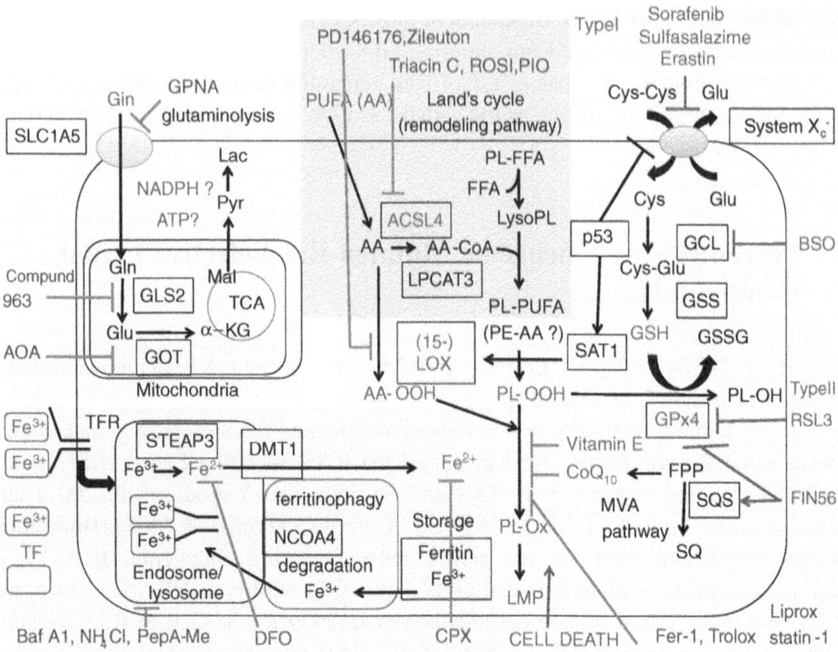

6.1 *Inhibition of System X_c^- Leads to Ferroptosis (Type I)*

System X_c^- is a membrane Na^+-dependent cysteine-glutamate exchange transporter, which is a disulfide-linked heterodimer composed of a light-chain subunit (xCT, SLC7A11) and a heavy-chain subunit (CD98hc, SLC3A2) (Sato et al. 1999). While system X_c^- transports intracellular glutamate to the extracellular space, it transports extracellular cystine into cells, which is the transformed into cysteine for glutathione (GSH) synthesis. Erastin acts as a direct inhibitor of system X_c^- function (Dixon et al. 2012) (Fig. 3). Erastin treatment leads to significant depletion of intracellular glutathione by inhibition of cellular uptake of cystine, since cystine is a key molecule for glutathione (GSH) synthesis. This effect could be inhibited by β-mercaptoethanol, because β-ME can enhance cysteine uptake through other pathways. Glutamate–cysteine ligase (γ-glutamylcyteine synthetase, GCL) is the rate-limiting first enzyme in the two step synthesis of glutathione (GSH). Blockage of GCL by buthionine-(S,R)-sulfoximine (BSO) can induce depletion of GSH, leading to ferroptosis that is prevented by vitamin E, Ferostatin-1, and DFO. These results indicated that inhibition of GSH synthesis is required to trigger ferroptosis in some cells (Seiler et al. 2008; Angeli et al. 2014). GPx4 can reduce direct toxic lipid peroxide (PLOOH) to nontoxic lipid alcohols (PL-OH) using GSH as a cofactor. Depletion of intracellular glutathione by erastin induced accumulation of lipid peroxidation detected by 2′,7′-dichlorodihydrofluorescein diacetate (DCFDA) and BODIPY-C11. Therefore, inhibition of system X_c^- by erastin suppresses GPx4

◂**Fig. 4** Regulation of ferroptosis by oncogenic mutated Ras-Selective Lethal compounds. Ferroptosis is an iron-dependent form of cell death involving the generation of phospholipid peroxidation (PL-Ox) in Ras-mutated cancer cells induced by oncogenic mutated Ras-selective lethal compounds (Type I and Type II RSL) as shown in pink zone. In cancer cells, cystine (Cys-Cys) uptake for maintaining of intracellular glutathione (GSH) via system X_c^- (pink zone), Iron uptake for production of iron containing enzymes, mitochondrial enzymes, p450 and lipoxygenase (LOX) via transferrin (TF) receptor (TFR) (yellow zone), and Glutamine (Gln) uptake for glutaminolysis in metabolic changes for energy production via SLC1A5 (green zone) were enhanced as compared to normal cells. Glutathione peroxidase 4 (GPx4) could reduce phospholipid hydroperoxide (PLOOH) to hydroxyl phospholipid (PLOH), oxidizing GSH to glutathione disulfide (GSSG) in the process. GPx4 is a key regulator of ferroptosis, since loss of GPx4 activity accumulates ferrous iron (Fe^{2+})-dependent oxidized phospholipid (PL-Ox), leading to the loss of membrane permeability (LMP) and cell death. Type I ferroptosis inducer, erastin, sulfasalazine and sorafenib, inhibit the cystine transportor, system X_c^-, resulting in the decrease of GSH, loss of GPx4 activity to ferroptosis. Type II ferroptosis inducer, RSL3 and FIN56 directly bound to GPx4 or induce the degradation of GPx4, resulting in iron-dependent phospholipid peroxidation to ferroptosis without the decrease of GSH. Ferroptosis was inhibited by iron-chelator DFO (deferoxiamine) and CPX (ciclopirox), and anti-lipid peroxidation compounds such as vitamin E, coenzyme Q10 (coQ10), Trolox and ferroptosis specific inhibitor, ferrostatin-1(Fer-1) and liproxstatin-1 (pink zone). Intracellular Fe^{2+} availability was regulated by degradation of extracellular transferrin (TF) bound Fe^{3+} incorporated by transferrin receptor (TFR) and intracellular ferritin bound Fe^{3+} in endosome / lysosome containing ferritinophagy (yellow zone). Sensitivity of phospholipid peroxidation in membrane is regulated by the quality of lipid membrane (Lipoquality) that means the change of the content of polyunsaturated fatty acid (PUFA) such as arachidonic acid (AA) in phospholipid membrane (violet zone). Acyl-CoA synthetase long-chain family member 4 (ACSL4) and lysophosphatidylcholine acyltransferase 3 (LPCAT3) could modulate the content of AA in phospholipid by remodeling pathway (Lands' pathway). The decrease of PL-PUFA is correlated to the sensitivity of lipid peroxidation and ferroptosis. Ferroptosis is required for transferrin and glutamine (Gln) in the culture media. Glutamine produces NADPH and ATP during glutaminolysis pathway, but this requirement of glutaminolysis remained to be resolved (green zone). p53 suppress the transcription of system X_c^- transportor and induction of spermidine/spermine *N*-acetyltransferase 1 (SAT1), up-regulator of 15-lipoxygenase (15-LOX). Other small molecule inducers of ferroptosis are indicated in red, while suppressors of ferroptosis are in blue. GCL Glutamate cysteine ligase, GSS glutathione synthetase, Cys, cysteine, Glu glutamate, Gln glutamine, α-KG alpha-ketoglutarate, GPNA l-g-glutamyl-p-nirtroanilide, AOA amino-oxyacetate, GLS2 glutaminase, GOT glutamine oxaloacetate aminotransferase, TCA tricarboxylic acd, Mal Malic acid, Pyr pyruvic acid, Lac lactic acid, DFO deferoxamine, CPX ciclopirox. Baf.A1 Bafiromycin A, PepA-Me pepstatin A-methyl ester, STEAP3 six-transmembrane epithelial antigen of the protease 3, DMT1 divalent metal transporter 1, NCOA4 cargo receptor for ferritinophagy, ROSI rosiglitazone, PIO pioglitazone, BSO buthione-(S,R)-sulfoximine, MVA Mevalonate, FPP Farnesyl diphosphate, SQS squalene synthase, SQ squalene

activity to cause accumulation of lethal lipid peroxides and to initiate the execution of ferroptosis. Erastin, sulfasalazine (Gout et al. 2001; Skouta et al. 2014) and sorafenib (Louandre et al. 2013; Lachaier et al. 2014) are ferroptosis inducers through this mechanism.

On the other hand, erastin is also reported to bind to voltage-dependent anion channels (VDAC2 and VDAC3) on the mitochondria to alter membrane permeability and the ion selectivity of the channels, leading to the mitochondrial dysfunction and oxidant release in ferroptosis (Yagoda et al. 2007). Erastin also leads to activation of an endoplasmic reticulum stress response and up-regulation of CHAC1 (cation transporter homolog 1) during ferroptosis (Dioxn et al. 2014). However, this relationship between these events in mitochondria and ER and ferroptosis remained to be elucidated.

The p53 protein has been well characterized for its response to various cellular stresses, including of growth arrest, senescence and apoptosis. p53 inhibits cystine uptake and sensitizes cells to ferroptosis by repressing expression of SLC7A11, a key component of the system X_c^- transporter with oxidants. p53 binds to the SLC7A11 locus at a specific p53 response element within the 5′ untranslated region (Jiang et al. 2015). p53 (3KR, R117, R161, and R162), an acetylation-defective mutant that fails to induce cell-cycle arrest, senescence and apoptosis, fully retains the ability to regulate SLC7A11 expression and induce ferroptosis upon reactive oxygen species (ROS)-induced stress. Acetylation of K98 of p53 is required for repression of transcription of SLC7A11 and induction of ferroptosis (Wang et al. 2016). p53 (4KR, R98, R117, R161, R162) could not induce tumor suppression. These results involve a new mode of tumor suppression based on p53 regulation of cystine metabolism, ROS responses and ferroptosis.

Addition of high concentration of Glu in neuronal cell lines induces the inactivation of system X_c^- and inhibition of uptake of cystine, leading to the GSH depletion and oxidative cell death (oxidative glutamate toxicity) named "Oxytosis" (Tan et al. 2001). Oxidative glutamate toxicity can be blocked by Ferrostatin-1, indicating that this cell death is ferroptosis (Dixon et al. 2012; Liu et al. 2015; Kang et al. 2014). GPx4 can suppress the glutamate-induced oxytosis in the retina (Sakai et al. 2015a). However, several reports indicated that mechanism of downstream of ROS accumulation in oxidative glutamate toxicity was different from ferroptosis (Henke et al. 2013; Tobaben et al. 2011). For example, extracellular Ca^{2+} influx, BH3 interacting domain death agonist (Bid) mediated mitochondrial damage and nuclear translocation of apoptosis inducing factor (AIF) are required for oxidative glutamate toxicity, but not required for ferroptosis. These differences remain to be resolved.

6.2 *Inhibition of Chemical Inactivation of GPx4 Leads to Ferroptosis (Type II)*

Type II inhibitor of ferroptosis such as (1S,3R)-RSL3 could inhibit ferroptosis without the decrease of GSH, whereas Ferrostatin-1, vitamin E, and DFO suppressed RSL3 induced ferroptosis (Yang and Stockwell 2008) (Fig. 4). GPx4 was identified as a major target protein of (1S,3R)-RSL3 by affinity purification with RSL3 and identification of sequence of amino acids. (1S,3R)-RSL3 contains an electrophilic chloroacetamide and covalently interacts with the active site "selenocysteine" of GPx4 to inhibit its enzymatic activity (Yang et al. 2014, 2016). Treatment of RSL3 induced the peroxidation of phospholipid that is prevented by vitamin E, ferrostatin-1, and DFO. GPx4 overexpression suppresses the RSL3-induced ferroptosis, whereas GPx4 knockdown enhances the sensitivity for RSL3-induced ferroptosis, suggesting that inhibition of GPx4 activity is a major contributor to ferroptosis. Thus, GPx4 is currently believed to be a key regulator in ferroptosis induced by erastin and RSL3.

Another screening of caspase-independent lethal compounds revealed new six specific ferroptosis inducers (Shimada et al. 2016). FIN56, one of six ferroptosis inducer, induced cell death was prevented by vitamin E, DFO, and U0126, MEK inhibitor in Ras-mutated cancer cell lines (BJelR). FIN56 treatment induced cell death for 48 h slowly that RSL3 at 8 h. But FIN56 treatment produces lipid peroxidation 3 h, whereas RSL3 produces it 1 h. FIN56 could not bind to GPx4. But FIN56 promoted degradation of GPx4, whereas a proteasome inhibitor MG132 did not inhibit FIN56 -induced cell death. 5-(tetradecyloxy)-2-furoic Acid (TOFA), an inhibitor of acetyl-CoA carboxylase (ACC), inhibited the degradation of GPx4. TOFA also suppresses FIN56-induced cell death and lipid peroxidation. ACC is an enzyme involved in fatty acid synthesis. But ACC itself was not direct target of FIN56. Chemoproteomics analysis revealed that FIN56 binds and activates squalene synthase (SQS), an enzyme involved in the cholesterol synthesis, in a manner independent of GPx4 degradation. Squalene synthase inhibitors blocked FIN56-induced ferroptosis and increased the mevalonate metabolite such as farnesyl pyrophosphate (FPP) and coenzyme Q_{10} (CoQ_{10}), an electron carrier in the mitochondrial respiratory chain and endogenous antioxidant. Supplementation of FPP or CoQ_{10} suppresses the lethality of FIN56 and RSL3. These results indicate that mechanism of cell death by FIN56 involves two distinct pathways, GPx4 degradation pathway that requires the activity of ACC, and activation of SQS, that leads to CoQ_{10} depletion as antioxidants. GPx4 degradation was also observed in RSL3-induced ferroptosis in MEF cells, indicating that RSL3 might involove another target protein except for GPx4 like FIN56 (Kagan et al. 2016; Shimada et al. 2016) (Fig. 4).

6.3 The Role of Iron Homeostasis in Ferroptosis

Iron is essential for the execution of ferroptosis. Erastin- and RSL3-induced ferroptosis can be inhibited by iron chelator of membrane impermeable (DFO) and membrane permeable (CPX) (Dixon et al. 2012; Yang et al. 2014).

Nutrient starvation in growth medium containing glucose but lacking both amino acids and serum, induces apoptosis in mouse embryonic fibroblasts (MEFs), whereas nutrient starvation medium in the presence of serum change from the apoptotic cell death to ferroptosis. Cystine deficiency of medium in the presence of serum causes ferroptosis in MEF cells, leading to the decrease of intracellular glutathione. This is interesting system that can change apoptosis to ferroptosis with or without serum (Gao et al. 2015). Two serum factors, the iron-carrier protein transferrin and amino acid glutamine, were identified as the inducer of ferroptosis by cystine deficiency and erastin. Ferroptosis by erastin or cystine deficiency is inhibited by genetic silence of transferrin receptor that required for the uptake of transferrin-iron complexes into cells (Gao et al. 2015; Yang and Stockwell 2008). Supplementing the medium with iron-bound transferrin or a bioavailable form of iron (ferric ammonium citrate) accelerates erastin-induced ferroptosis (Gao et al. 2015; Dixon et al. 2012) and GPx4-depleted cell death (Sakai et al. 2016a). These results confirmed that the requirement for iron in ferroptosis (Fig. 4).

L-glutamine (L-Gln) is the most abundant amino acid in the body. Proliferating cells use L-Gln both as a nitrogen source for the biosynthesis of nucleotides, amino acids, and hexamine and as an important carbon source for the tricarboxylic acid (TCA) cycle. Pharmacological inhibition of SLC1A5, L-Gln receptor component by L-g-glutamyl-p-nirtroanilide (GPNA) or RNAi knockdown of these receptor prevented ferroptosis induced by cystine deficiency. Gln is converted into glutamate (Glu) by glutaminases (GLS). An inhibitor of GLS, Compound 968, and RNAi knockdown of GLS2 but not GLS1 blocked ferroptosis induced by erastin or cystine deficiency. Downstream of glutaminolysis, glutamate can be further converted into a-ketoglutarate (a-KG) either transaminase-mediated transamination or by glutamate dehydrogenase (GLUD1)-mediated glutamate deamination. The amino-oxyacetate (AOA), a pan inhibitor of transaminases and RNAi of transaminase GOT1 inhibited ferroptosis induced by erastin and cystine deficiency, but not GLUD1 RNAi could not suppress. The glutamine-fueled intracellular metabolic pathway, glutaminolysis, involved important roles in ferroptosis in cancer cells. Inhibition of glutaminolysis can reduce heart injury triggered by ischemia/reperfusion. These results indicated that glutaminolysis is essential for ferroptosis. NADPH and ATP, which are produced by glutaminolysis might be related to the execution of ferroptosis in cancer cells (Fig. 4).

The Ras-mutated cancer cells accumulate iron by modulating expression levels of ferritin and the transferrin receptor (Yagoda et al. 2007). Iron that is taken up by the transferrin receptor bound to transferrin-Fe(III)$_2$ is routed into the endosome/lysosome pathway. The ferric ion, Fe(III), that is released from transferrin is reduced by an endosomal reductase activity (e.g., six-transmembrane epithelial antigen of the prostate 3) prior to export of ferrous ion, Fe(II), by a transporter such as divalent metal transporter1 (DMT1). Most cellular iron is stored as a component of ferritin in the Fe (III) form (Hentze et al. 2010). Recent studies showed that selective autophagic degradation of ferritin by the cargo receptor NCOA4 that recognized ferritin promotes iron release within lysosomes, and this ferritinophagy is important for cellular iron homeostasis for the increase of free iron (Mancias et al. 2014; Dowdle et al. 2014; Bellelli et al. 2016). The cellular labile iron pool might be used to control intracellular iron homeostasis through its own redox activity or to provide iron constituents for functional proteins such as heme enzymes including P450, cyclooxygenase, NADPH oxidases (NOXs), and non-heme enzyme including lipoxygenase.

Bafiromycin A1 (Baf A1), an inhibitor of vacuolar H^+-ATPase, the lysosomal aspartic protease inhibitor pepstatin A-methyl ester (PepA-Me), ammonium chloride (NH_4Cl) that acts to neutralize acidic organelle such as lysosome blocked erastin and RSL3-induced ferroptosis and ROS production. PepA-Me caused both an increase in ferritin protein levels and a decrease in iron content, indicating this may block ferroptosis by preventing autophagic degradation of ferritin, whereas Baf A1 and NH_4Cl inhibit the iron uptake by transferrin (Torii et al. 2016).

Inhibition of ferritinophagy by prevention of autophagy such as ATG13 and ATG3 or knockdown of cargo receptor NCOA4 for ferritinophagy, abrogated the accumulation of ferroptosis-associated cellular labile iron and ROS as well as ferroptosis. These results indicated endosomes/lysosomes and NCOA4 mediated ferritinophagy contribute to ferroptosis by modulating cellular iron homeostasis and ROS production (Gao et al. 2016; Hou et al. 2016) (Fig. 4).

Screening of shRNA suppressors of erastin in U2-OS cells revealed that PHKG2, the catalytic subunit of the PHK (phosphorylase kinase) complex, that activates glycogen phosphorylase (GP) to release glucose-1-phosphate from glycogen. PHK inhibitor, K252a inhibit erastin-induced ferroptosis, but not two glycogen phosphorylase inhibitors could not prevent, indicating that the metabolic pathway of glycogen breakdown is not responsible for the suppression of erastin-induced ferroptosis. Knockdown of PHKG2 suppressed the production of ROS and the availability of ferrous iron, indicating that novel iron regulatory function of PHKG2 is responsible for modulating sensitivity to erastin (Yang et al. 2016). Thus, iron availability to lipoxygenase by PHKG2 and lysosomal ferritin degradation is essential for erastin-induced ferroptosis.

Other pathways regulating the iron availability are also involved in the modulation of ferroptosis. Heat shock protein HSPB1 inhibition or heme oxygenase-1 (HO-1) induction enhanced the sensitivity of erastin-induced ferroptosis by the increase of intracellular free iron (Sun et al. 2015; Kwon et al. 2015).

6.4 Regulation of Phospholipid Peroxidation Modulate the Sensitivity of Ferroptosis

Lipid peroxidation occurs through the mechanism of both an enzymatic lipoxygenase reaction and a non-enzymatic free-radical chain reaction by Fenton reactions. Fenton reaction is defined as the oxidation of organic substrates by a mixture of hydrogen peroxide and ferrous iron. The Fenton reaction is a key reaction in oxidation of phospholipids and no specificity for classes of phospholipids. Singlet molecular oxygen also participates in a non-enzymatic reaction, which yields lipid hydroperoxide (LOOHs) from cholesterol and esterified lipids such as phospholipids and cholesteryl esters (Hauck and Bernlohr 2016). Ferroptosis is an iron-dependent cell death, accumulating lipid peroxidation. GPx4 that could reduce phospholipid hydroperoxide is a regulator of ferroptosis. Vitamin E or Ferrostatin-1 could suppress peroxidation of phospholipid and cell death in ferroptosis. The current proposal model is that after GPx4 activity is lost, lipid hydroperoxide was generated in membrane and attacked with ferrous iron, resulting in formation of lipid radicals and enhancement of lipid peroxidation to the lethal damage of membrane. Erastin and RSL3 treatment resulted in accumulation of several classes of phospholipid hydroperoxide such as phosphatidylcholine (PC), phosphatidylethanolamine (PE), phosphatidylserine (PS) and lysophospholipids and depletion of several polyunsaturated fatty acid (PUFA) such as arachidonic acid (AA, 20:4n-6), indicating that specific oxidized PUFA are cleaved from the glycerophospholipids by phospholipase A_2 and subsequently degraded or destroyed in plasma membrane permeability (Angeli et al. 2014; Dixon et al. 2015; Yang et al. 2016) (Fig. 4).

Recent reports demonstrated that depletion of acyl-CoA synthetase long-chain family member 4 (ACSL4) or lysophophatidylcholine acyltransferase 3 (LPCAT3) that encode enzymes required for the reacylation of membrane lysophospholipid such as lyso PC (lysophosphatidylcholine) and lysoPE (Lysophosphatidylethanolamine) with arachidonic acid and other PUFAs (Hashidate-Yoshida et al. 2015), inhibits ferroptosis induced by RSL3 and erastin (Dixon et al. 2015; Doll et al. 2016; Kagan et al. 2016; Yuan et al. 2016). Cell membrane contains several classes of glycerophospholipids, which have numerous structural and functional roles in the cells. Polyunsaturated fatty acids, including arachidonic acid, eicosapentanoic acid, docosahexanoic acid, are located at the sn-2 (but not sn-1)-position of glycerophospholipids in an asymmetrical manner. Using acyl-CoAs as donors, glycerophospholipids are formed by a de novo pathway (Kennedy pathway) and modified by a remodeling pathway (Lands' pathway) to generate membrane asymmetry and diversity (Sindou and Shimizu 2009). ACSL4 has a marked preference for long-chain PUFAs such as arachidonic acid (AA) (Yan et al. 2015; Cho et al. 2001). The content of PUFA in biomembrane such as phosphatidylcholine (PC) and phosphatidylethanolamine (PE) were complicatedly regulated by lysophophatidylcholine acyltransferase (LPCAT)1-4 and lysophophatidylethanolamine acyltransferase (LPEAT)1-2, remodeling enzymes and phospholipase A_2. The remodeling of arachidonic acids in PC, PE, and PI were controlled by acyl-CoA synthetase long-chain family member 4 (ACSL4) and LPCAT2, LPEAT 2, LPCAT3 and LPIAT (Fig. 4).

Conrad' group showed that ACSL4 KO cells are significantly resistant for RSL3 induced ferroptosis in Pfa1 cells (MEF cells), whereas only a mild protective effect was observed in LPCAT3 KO cells. Loss of ACSL4 gene inhibited RSL3 induced and tamoxifen inducible GPx4 depleted induced ferroptosis and peroxidation of phospholipid. Decrease of arachidonic acid (AA: 20:4) and adrenoyl acid (AdA: 22:4) containing PE (phosphatidylethanolamine) and PI (phosphatidylinositol) were markedly less abundant in ACSL4 KO cells than in WT cells, whereas no decrease was found in phosphatidylcholine and phosphatidylserine (Doll et al. 2016). Kagan' group showed that loss of ACLS4 suppressed the formation of double and triple-oxidized AA– and AdA containing PE species (15–OOH–AA–PE, 15–OOH–8OH–AA–PE, 15–OOH–9OH–AA–PE and 15–OOH–12OH–PE) in RSL3-induced ferroptosis, whereas the elevation of four oxidized PE was enhanced in RSL3 induced WT cells (Pfa1 cells) and kidneys of GPx4KO mice. Lipoxygenase inhibitor (NCTT-956, PD146176, ML351, and Zileuton) could suppress the RSL3 induced ferroptosis in Pfa1 cells, but not COX inhibitor (Piroxicam) and P450 inhibitor (MSPPOH) could not. Liperfluo fluorescence intensity that was detected for phospholipid hydroperoxide was elevated in endoplasmic reticulum in RSL3-induced ferroptosis. His group demonstrated that oxidation of AA or AdA containing PE by 15-lipoxygenase (15-LOX) in ER is required for the execution for RSL3-induced ferroptosis in Pf1a cells as a ferroptotic death signal (Kagan et al.). Lipidomics analysis by other groups showed that phosphatidylcholine with PUFA was depleted, whereas the levels of ceramide and lysoPC (lysophosphatidylcholine) accumulated during erastin-induced ferroptosis in HT-1080 fibrosarcoma cells. In G401 cells and HT1080 cells, 15-LOX knockdown could suppress the erastin-induced ferroptosis, but not RSL3-induced ferroptosis. Also lipoxygenase inhibitor, Zileuton, Baicalein, and PD146176 could suppress the erastin-induced ferroptosis in G401 cells, but not cyclooxygenase (COX) inhibitor indomethacin (Dixon et al. 2015). ACSL4 knockdown could suppress erastin induced ferroptosis in HepG2 and HL60 cells. ACSL4 overexpression could recover erastin-induced ferroptosis in no ACSL4 expressing erastin-resistant cell lines, LNCaP and K562 cells. Addition of 5-lipoxygenase inhibitor, Zileuton could suppress the erastin-induced ferroptosis in LNCaP cells (Yuan et al. 2016). In these cell lines, 5-lipoxygenase is important for erastin-induced ferroptosis. GPx4 and vitamin E directly could suppress the initial activation of lipoxygenase by lipid hydroperoxide as an activator (Imai et al. 1998). These results indicated that 15-LOX is one of the candidate for initial oxidation of phospholipid in RSL3 and erastin-induced ferroptosis (Fig. 4).

The emerging role of p53 in ferroptosis has been a topic of great interest. But it is unclear how p53 orchestrates its activities in multiple metabolic pathways into tumor suppressive effects (Jiang et al. 2015; Wang et al. 2016). Spermidine/spermine N^1-acetyltransferase 1 (SAT1) gene was identified as a transcription target molecule of p53. SAT1 is a rate-limiting enzyme in polyamine catabolism. Interestingly induction of SAT1 mRNA results in the lipid peroxidation and ferroptosis upon reactive oxygen species induced stress. SAT1 induction was correlated with induction of 15-LOX mRNA. SAT1-induced ferroptosis significantly suppressed in the presence of PD146176, specific inhibitor of 15-LOX. Thus, 15-LOX is the key regulator of ferroptosis induced by SAT1, p53 up-regulator molecule (Ou et al. 2016).

On the other hand, in 15-LOX knockout MEF cells, GPx4 depletion and erastin also could induce ferroptosis. Expression of 15-LOX is especially localized in inflammatory cells and cancer cells. GPx4 depletion in 15-LOX deficient MEF cells also could induce the lipid peroxidation-dependent cell death (Angeli et al. 2014).

These evidences demonstrated that 15-LOX is not essential for the execution for ferroptosis and other oxidation system might be required for lipid oxidation during ferroptosis in 15-LOX deficient cells.

ACSL4 was preferentially expressed in a panel of basal-like breast cancer cell lines and its expression appeared to be strongly correlated with sensitivity to ferroptosis induction by RSL3. GPx4-Acsl4 double knockout cells showed marked resistance to ferroptosis, however RSL3 induced ferroptosis in GPx4-Acsl4 double knockout cells by supplementation of arachidonic acid. Interestingly GPx4-Acsl4 double knockout MEF cells were viable and proliferated normally in cell culture. But we showed that GPx4 knockout MEF cells also was viable and proliferated normally in cell culture with vitamin E. From these results, we proposed that antioxidant balance of lipid peroxidation in phospholipid such as elevation of vitamin E and decrease of PUFA in phospholipid might regulate the fate of cells (Fig. 3). And disruption of imbalance of oxidation of phospholipid might cause lipid peroxidation-dependent cell death such as ferroptosis.

7 Lipid Peroxidation-Dependent Cell Death by GPx4 Gene Disruption

GPx4 depletion by Cre-LoxP system or knockdown strategy induced lipid peroxidation-dependent cell death in many cells such as T cells (Matsushita et al. 2015), corneal endothelial cells (Uchida et al. 2016), conjunctival epithelial cells (Sakai et al. 2015b), vascular endothelial cells (Sakai et al. 2016b) and keratinocytes (Sengupta et al. 2013). These cell death without anti-cancer drug such as erastin and RSL3 was also rescued by vitamin E or ferrostain-1, anti-lipid peroxidation compounds.

Mechanism of cell death by depletion of GPx4 gene were mainly reported using tamoxifen inducible GPx4 depleted MEF cells (Pfa1 cells) from GPx4 flox/flox mice established by Conrad' group (Seiler et al. 2008; Angeli et al. 2014; Doll et al. 2016; Kagan et al. 2016). Pfa1 cells have the expression of 15-lipoxygenase (15-LOX). Tamoxifen inducible GPx4 depleted cell death in Pfa1 cells was inhibited by 15-LOX inhibitor, iron-chelator deferoxamine, ferrostatin-1 and vitamin E. RSL3, a direct inhibitor of GPx4, induced ferroptosis in Pfa1 cells also were inhibited by 15-LOX inhibitor, Baicalen, PD146176, and ACSL4 inhibitor, Triacsin C, pioglitazone, iron-chelator deferoxamine, ferrostatin-1 and vitamin E. Active site of 15-LOX contain ferrous irons and GPx4 inhibit the activation of 15-LOX. From these results, Kagan et al. demonstrated that GPx4 depletion in Pfa1 cells initiate 15-Lipoxygenase activation dependent phospholipid peroxidation and induced ferroptosis (Kagan et al. 2016; Doll et al. 2016).

Until now, the mechanism of ferroptosis by both RSL-induced and GPx4 depleted cell death by GPx4 gene disruption in MEF cells are considered to be the same cell death mechanism. However, our established tamoxifen inducible GPx4 depleted MEF cells (ETK cells) did not express 15-LOX, 12-LOX, and 5-LOX (Imai et al. 2009; Imai 2010). In ETK cells, GPx4 depleted cell death by tamoxifen treatment is not inhibited by apoptosis inhibitor Z-BAD-FMK, by knockdown of autophagy regulating protein ATG5 and necroptosis regulator RIP kinase 1, indicating that GPx4 depleted cell death is caspase-independent non-apoptotic cell death. In ETK cells, erastin and RSL3 also induced ferroptosis as reported previously, since these cell deaths could inhibit by iron chelator, ferrostatin-1 and vitamin E. However, GPx4 depleted cell death by tamoxifen treatment in ETK cells could not be inhibited by 15-LOX inhibitor, Baicalen and ACSL4 inhibitor, Triacsin C and pioglitazone, iron-chelator deferoxamine, whereas ferrostatin-1 and vitamin E effectively suppressed the cell death. In ETK cells, erastin and RSL3 induce ferroptosis by 12 h, but tamoxifen treatment induced GPx4 depleted cell death 72–96 h after treatment although GPx4 expression could not be detected by 24 h. Lipid peroxidation was detected 6h after treatment of erastin in ETK cells and suppressed by iron chelator. However, lipid peroxidation by tamoxifen treatment was observed at 26 h early time before cell death at 72 h. Lipid peroxidation 26 h after tamoxifen treatment was suppressed by vitamin E, but not by iron chelator. Addition of scavengers for superoxide, hydrogen peroxide, and overexpression of antioxidant enzyme such as SOD1, SOD2, and GPx1 could not rescue GPx4-depleted cell death in ETK cells (Imai 2010). When vitamin E is added by 26 h after treatment of tamoxifen, GPx4-depleted cell death can be effectively inhibited, but the lethality can not be suppressed after 26 h, indicating iron-independent lipid peroxidation by 26 h is required for GPx4 cell death in ETK cells. Short hairpin RNA (shRNA) mediated knockdown of GPx4 in 15-LOX null-MEF cells is sufficient to induce cell death, and vitamin E also inhibited GPx4 deleted cell death in 15-LOX null-MEF cells. These results indicated that 15-LOX, independent and iron-independent lipid peroxidation is necessary for GPx4-depleted cell death in ETK cells.

Although RSL3 and erastin quickly induced cell death 12 h in ETK cells, GPx4 depletion by tamoxifen slowly induced cell death 72–96 h in ETK cells. One possibility of the differences of time for cell death between GPx4 gene disruption and RSL-compounds is that the existence of another accelerate pathway in ferroptosis by RSL except for GPx4 inactivation, since erastin can bind to VDAC in mitochondria and induce ER stress and RSL can bind to the several proteins except for GPx4. Recent works demonstrated that RSL3 could degrades GPx4 in MEF cells (Pfa1 cells) like FIN56, however whether protein degradation by RSL3 is specific for GPx4 or not remained to be solved (Kagan et al. 2016; Shimada et al. 2016). The other possibility is that the mechanism of GPx4 depleted cell death by GPx4 gene disruption is different from the mechanism of ferroptosis by erastin and RSL3.

In mouse erythroid precursor cells, inactivation of GPx4 induced to the accumulation of toxic lipid intermediates that covalently modify caspase-8 and trigger necroptosis in the absence of death receptor stimulation such as TNFα (Canli et al. 2016). GPx4 inhibition can sensitize cancer cells to apoptosis induced by second mitochondrial-derived activator of caspase (SMAC) mimetics, connecting ferroptosis to apoptotic cell death pathway (Dächert et al. 2016). These results demonstrated that differences of lipid peroxidation producing system such as specific enzyme and random oxidation by Fenton reaction, and producing local site in organelle in cells might execute different cell death signaling, such as apoptosis, ferroptosis, necroptosis, and novel lipid peroxidation cell death.

Thus, GPx4 could regulate several cell death pathways by suppression of phospholipid peroxidation in specific site or local site of organelle (Fig. 2).

8 Ferroptosis in Disease Model

Analysis of role of ferroptosis in pathological cell death has been enabled by the ferroptosis specific small molecule, Ferrostatin-1 (Fer-1) (Dixon et al. 2012), Liproxstatin-1 (Angeli et al. 2014) and vitamin E.

Erastin enhanced chemotherapy drug such as temozolomide, cisplatin, cytarabine/ara-C, and doxorubicin/Adriamycin in certain cancer cells (Yu et al. 2015; Chen et al. 2015; Yamaguchi et al. 2013). In vivo, erastin, piperazine erastin, and RSL3 inhibited tumor growth in a xenograft model (Yang et al. 2014; Sun et al. 2015). In a rat organotypic hippocampal slice culture model, glutamate-induced neurotoxicity was prevented by Fer-1 (Dixon et al. 2012). Fms-like tyrosine kinase3 (FLT-3, also termed CD135) is a cytokine receptor, that is important for the normal development of hematopoietic stem cells and progenitor cells. Inhibitors for Fms-like tyrosine kinase3 (FLT-3) and its downstream signaling molecule phosphoinositide 3-kinase α can suppress lipid peroxidation to inhibit ferroptosis in neuron (Kang et al. 2014).

In a Huntington's disease model, Fer-1 restored the number of healthy neurons by inhibition of ferroptosis (Skouta et al. 2014). Fer-1 significantly protected the death of developing oligodendrocytes from cystine deprivation (Skouta et al. 2014).

Fer-1 prevented lethality in a model of acute injury of freshly isolated renal tubules, implicating ferroptosis-mediated cell death by acute kidney failure (Skouta et al. 2014). Ferrostatin analog (SRS 16–86) inhibits acute ischemia-reperfusion injury and oxalate nephropathy related acute kidney failure (Linkermann et al. 2014). Inducible knockout of GPx4 in the kidney leads to ferroptosis, which contributes to acute kidney failure in mice (Angeli et al. 2014).

High dose of acetaminophen frequently cause acute liver failure. Fer-1 can inhibit acetaminophen induced ferroptotic cell death (Lorincz et al. 2015). Ischemia/reperfusion-induced liver injury can be prevented in mice by liproxstatin-1 (Angeli et al. 2014). Prevention of glutaminolysis and ferroptosis by compound 968, DFO or Fer-1 inhibits ischemia/reperfusion-induced heart injury ex vivo (Gao et al. 2015).

9 Conclusion and Prospective

Ferroptosis by erastin and RSL3 is an iron-dependent lipid peroxidation induced non-apoptotic cell death in RAS-mutated cancer cells. Erastin inhibits cystine transporter activity and induces the decrease of glutathione and GPx4 activity, resulting in iron or 15-LOX dependent lipid peroxidation induced cell death. 15-LOX is one of the candidates for initial lipid peroxidation in ferroptosis. Fenton reaction by ferrous iron enhances the propagation of phospholipid oxidation and degradation of membrane lipid. Ferroptosis inducer is important for therapy of cancer. Identification of the downstream signaling pathway or executors of 15-LOX independent lipid peroxidation in ferroptosis remained to be solved.

GPx4 could directly reduce phospholipid hydroperoxide in specific organelle and regulate several signal transductions by analysis of GPx4 overexpressing cells. Overexpression of mGPx4 inhibited apoptosis induced by mitochondrial death pathway. Overexpression of cGPx4 also suppressed the iron-dependent lipid peroxidation in membrane induced by erastin and RSL3, resulting in inhibition of ferroptosis. Where and how lipid peroxidation in organelle such as ER, Golgi, and plasma membrane is generated in ferroptosis by erastin and RSL3 remained to be elucidated.

Indeed, GPx4 is a regulator of ferroptosis by erastin and RSL3. But cell death by GPx4 gene disruption progresses extremely slower than ferroptosis induced by erastin and RSL3. It may be possible that the mechanism of cell death is different between ferroptosis by erastin and RSL3 and GPx4 depleted cell death by GPx4 gene disruption. Differences of site and enzymes of phospholipid oxidation in organelle might show differences of its downstream signal transduction of cell death between ferroptosis by erastin and RSL3 and GPx4 depleted cell death.

The phenotype of tissue specific GPx4 KO mice and cells is recovered with treatment of vitamin E. These results demonstrated that imbalance between lipid oxidation system and lipid peroxidation suppression system such as GPx4 and vitamin E causes several diseases in mice and human.

Acknowledgements We thank H. Nakano and S. Nagata for helpful comments on the manuscript. We also thank members of Department of Hygienic Chemistry, School of Pharmaceutical Sciences, Kitasato University for helpful discussion. This work was supported in part by Grants-in-Aid from Scientific Research (C) (26460075) from JSPS KAKENHI and Scientific Research on Innovative Areas (15H01386 and 16H01367) from a MEXT (Ministry of Education, Culture, Sports, Science and Technology), Japan, and research grants from Iijima Tojyuro Memorial Food Science Foundation and Kitasato University Research Grant for Young Researchers.

Competing Interests The authors declare that they have no competing interests.

References

Angeli JPF, Schneider M, Proneth B, Tyurina YY, Tyurin VA, Hammond VJ, Herbach N, Aichler M, Walch A, Eggenhofer E, Basavarajappa D, Rådmark O, Kobayashi S, Seibt T, Beck H, Neff F, Esposito I, Wanke R, Förster H, Yefremova O, Heinrichmeyer M, Bornkamm GW, Geissler EK, Thomas SB, Stockwell BR, O'Donnell VB, Kagan VE, Schick JA, Conrad M (2014) Inactivation of the ferroptosis regulator Gpx4 triggers acute renal failure in mice. Nat Cell Biol 16:1180–1191

Arai M, Imai H, Sumi D, Imanaka T, Takano T, Chiba N, Nakagawa Y (1996) Import into mitochondria of phospholipid hydroperoxide glutathione peroxidase requires a leader sequence. Biochem Biophys Res Commun 227:433–439

Arai M, Imai H, Koumura T, Yoshida M, Emoto K, Umeda M, Chiba N, Nakagawa Y (1999) Mitochondrial phospholipid hydroperoxide glutathione peroxidase plays a major role in preventing oxidative injury to cells. J Biol Chem 274:4924–4933

Bellelli R, Federico G, Matte' A, Colecchia D, Iolascon A, Chiariello M, Santoro M, De Franceschi L, Carlomagno F (2016) NCOA4 deficiency impairs systemic iron homeostasis. Cell Rep 14:411–421

Canli Ö, Alankuş YB, Grootjans S, Vegi N, Hültner L, Hoppe PS, Schroeder T, Vandenabeele P, Bornkamm GW, Greten FR (2016) Glutathione peroxidase 4 prevents necroptosis in mouse erythroid precursors. Blood 127:139–148

Carlson BA, Tobe R, Yefremova E, Tsuji PA, Hoffmann VJ, Schweizer U, Gladyshev VN, Hatfield DL, Conrad M (2016) Glutathione peroxidase 4 and vitamin E cooperatively prevent hepatocellular degeneration. Redox Biol 9:22–31

Chen L, Li X, Liu L, Yu B, Xue Y, Liu Y (2015) Erastin sensitizes glioblastoma cells to temozolomide by restraining xCT and cystathionine-gamma-lysae function. Oncol Rep 33:1465–1474

Cho YY, Kang MJ, Sone H, Suzuki T, Abe M, Igarashi M, Tokunaga T, Ogawa S, Takei YA, Miyazawa T, Sasano H, Fujino T, Yamamoto TT (2001) Abnormal uterus with polycysts, accumulation of uterine prostaglandins, and reduced fertility in mice heterozygous for acyl-CoA synthetase 4 deficiency. Biochem Biophys Res Commun 284:993–997

Conrad M, Moreno SG, Sinowatz F, Ursini F, Kölle S, Roveri A, Brielmeier M, Wurst W, Maiorino M, Bornkamm GW (2005) The nuclear form of phospholipid hydroperoxide glutathione peroxidase is a protein thiol peroxidase contributing to sperm chromatin stability. Mol Cell Biol 25:7637–7644

Dächert J, Schoeneberger H, Rohde K, Fulda S (2016) RSL3 and erastin differentially regulate redox signaling to promote Smac mimetic-induced cell death. Oncotarget. doi:10.18632/oncotarget.11687. (Epub ahead of print)

Dixon SJ, Lemberg KM, Lamprecht MR, Skouta R, Zaitsev EM, Gleason CE, Patel DN, Bauer AJ, Cantley AM, Yang WS, Morrison B III, Stockwell BR (2012) Ferroptosis: an iron-dependent form of nonapoptotic cell death. Cell 149:1060–1072

Dixon SJ, Patel DN, Welsch M, Skouta R, Lee ED, Hayano M, Thomas AG, Gleason CE, Tatonetti NP, Slusher BS, Stockwell BR (2014) Pharmacological inhibition of cystine-glutamate exchange induces endoplasmic reticulum stress and ferroptosis. Elife. 23: e02523

Dixon SJ, Winter GE, Musavi LS, Lee ED, Snijder B, Rebsamen M, Superti-Furga G, Stockwell BR (2015) Human haploid cell genetics reveals roles for lipid metabolism genes in nonapoptotic cell death. ACS Chem Biol 10:1604–1609

Doll S, Proneth B, Tyurina YY, Panzilius E, Kobayashi S, Ingold I, Irmler M, Beckers J, Aichler M, Walch A, Prokisch H, Trümbach D, Mao G, Qu F, Bayir H, Füllekrug J, Scheel CH, Wurst W, Schick JA, Kagan VE, Angeli JP, Conrad M (2016) ACSL4 dictates ferroptosis sensitivity by shaping cellular lipid composition. Nat Chem Biol. doi:10.1038/nchembio.2239. (Epub ahead of print)

Dowdle WE, Nyfeler B, Nagel J, Elling RA, Liu S, Triantafellow E, Menon S, Wang Z, Honda A, Pardee G, Cantwell J, Luu C, Cornella-Taracido I, Harrington E, Fekkes P, Lei H, Fang Q, Digan ME, Burdick D, Powers AF, Helliwell SB, D'Aquin S, Bastien J, Wang H, Wiederschain D, Kuerth J, Bergman P, Schwalb D, Thomas J, Ugwonali S, Harbinski F, Tallarico J, Wilson CJ, Myer VE, Porter JA, Bussiere DE, Finan PM, Labow MA, Mao X, Hamann LG, Manning BD, Valdez RA, Nicholson T, Schirle M, Knapp MS, Keaney EP, Murphy LO (2014) Selective VPS34 inhibitor blocks autophagy and uncovers a role for NCOA4 in ferritin degradation and iron homeostasis in vivo. Nat Cell Biol 16:1069–1079

Fujii J, Imai H (2014) Redox reactions in mammalian spermatogenesis and the potential targets of reactive oxygen species under oxidative stress. Spermatogenesis 4:e979108

Gao M, Monian P, Quadri N, Ramasamy R, Jiang X (2015) Glutaminolysis and transferrin regulate ferroptosis. Mol Cell 59:298–308

Gao M, Monian P, Pan Q, Zhang W, Xiang J, Jiang X (2016) Ferroptosis is an autophagic cell death process. Cell Res 26:1021–1032

Gout PW, Buckley AR, Simms CR, Bruchovsky N (2001) Sulfasalazine, a potent suppressor of lymphoma growth by inhibition of the x_c^- cystine transporter: a new action for an old drug. Leukemia 215:1633–1640

Hashidate-Yoshida T, Harayama T, Hishikawa D, Morimoto R, Hamano F, Tokuoka SM, Eto M, Tamura-Nakano M, Yanobu-Takanashi R, Mukumoto Y, Kiyonari H, Okamura T, Kita Y, Shindou H, Shimizu T (2015) Fatty acid remodeling by LPCAT3 enriches arachidonate in phospholipid membranes and regulates triglyceride transport. Elife. 4. doi:10.7554/eLife.06328

Hauck AK, Bernlohr DA (2016) Oxidative stress and lipotoxicity. J Lipid Res 57:1976–1986

Henke N, Albrecht P, Bouchachia I, Ryazantseva M, Knoll K, Lewerenz J, Kaznacheyeva E, Maher P, Methner A (2013) The plasma membrane channel ORAI1 mediates detrimental calcium influx caused by endogenous oxidative stress. Cell Death Dis 4:e470

Hentze MW, Muckenthaler MU, Galy B, Camaschella C (2010) Two to tango: regulation of mammalian iron metabolism. Cell 142:24–38

Hou W, Xie Y, Song X, Sun X, Lotze MT, Zeh HJ III, Kang R, Tang D (2016) Autophagy promotes ferroptosis by degradation of ferritin. Autophagy 12:1425–1428

Imai H (2010) New strategy of functional analysis of PHGPx knockout mice model using transgenic rescue method and Cre-LoxP system. J Clin Biochem Nutr 46:1–13

Imai H (2011) Disruption of homeostasis of suppression of phospholipid peroxidation in biomembrane case novel cell death in disease. Oleoscience 11:15–23

Imai H, Nakagawa Y (2003) Biological significance of phospholipid hydroperoxide glutathione peroxidase (PHGPx, GPx4) in mammalian cells. Free Radic Biol Med 34:145–169

Imai H, Sumi D, Hanamoto A, Arai M, Sugiyama A, Nakagawa Y (1995) Molecular cloning and functional expression of a cDNA for rat phospholipid hydroperoxide glutathione peroxidase: 3′-untranslated region of the gene is necessary for functional expression. J Biochem 118: 1061–1067

Imai H, Sumi D, Sakamoto H, Hanamoto A, Arai M, Chiba N, Nakagawa Y (1996) Overexpression of phospholipid hydroperoxide glutathione peroxidase suppressed cell death due to oxidative damage in rat basophile leukemia cells (RBL-2H3). Biochem Biophys Res Commun 222:432–438

Imai H, Narashima K, Arai M, Sakamoto H, Chiba N, Nakagawa Y (1998) Suppression of leukotriene formation in RBL-2H3 cells that overexpressed phospholipid hydroperoxide glutathione peroxidase. J Biol Chem 273:1990–1997

Imai H, Suzuki K, Ishizaka K, Ichinose S, Oshima H, Okayasu I, Emoto K, Umeda M, Nakagawa Y (2001) Failure of the expression of phospholipid hydroperoxide glutathione peroxidase in the spermatozoa of human infertile males. Biol Reprod 64:674–683

Imai H, Hirao F, Sakamoto T, Sekine K, Mizukura Y, Saito M, Kitamoto T, Hayasaka M, Hanaoka K, Nakagawa Y (2003a) Early embryonic lethality caused by targeted disruption of the mouse PHGPx gene. Biochem Biophys Res Commun 305:278–286

Imai H, Koumura T, Nakajima R, Nomura K, Nakagawa Y (2003b) Protection from inactivation of the adenine nucleotide translocator during hypoglycaemia-induced apoptosis by mitochondrial phospholipid hydroperoxide glutathione peroxidase. Biochem J 371:799–809

Imai H, Saito M, Kirai N, Hasegawa J, Konishi K, Hattori H, Nishimura M, Naito S, Nakagawa Y (2006) Identification of the positive regulatory and distinct core regions of promoters, and transcriptional regulation in three types of mouse phospholipid hydroperoxide glutathione peroxidase. J Biochem 140:573–590

Imai H, Hakkaku N, Iwamoto R, Suzuki J, Suzuki T, Tajima Y, Konishi K, Minami S, Ichinose S, Ishizaka K, Shioda S, Arata S, Nishimura M, Naito S, Nakagawa Y (2009) Depletion of selenoprotein GPx4 in spermatocytes causes male infertility in mice. J Biol Chem 284:32522–32532

Jiang L, Kon N, Li T, Wang SJ, Su T, Hibshoosh H, Baer R, Gu W (2015) Ferroptosis as a p53-mediated activity during tumour suppression. Nature 520:57–62

Kagan VE, Tyurin VA, Jiang J, Tyurina YY, Ritov VB, Amoscato AA, Osipov AN, Belikova NA, Kapralov AA, Kini V, Vlasova II, Zhao Q, Zou M, Di P, Svistunenko DA, Kurnikov IV, Borisenko GG (2005) Cytochrome c acts as a cardiolipin oxygenase required for release of proapoptotic factors. Nat Chem Biol 1:223–232

Kagan VE, Mao G, Qu F, Angeli JP, Doll S, Croix CS, Dar HH, Liu B, Tyurin VA, Ritov VB, Kapralov AA, Amoscato AA, Jiang J, Anthonymuthu T, Mohammadyani D, Yang Q, Proneth B, Klein-Seetharaman J, Watkins S, Bahar I, Greenberger J, Mallampalli RK, Stockwell BR, Tyurina YY, Conrad M, Bayır H (2016) Oxidized arachidonic and adrenic PEs navigate cells to ferroptosis. Nat Chem Biol. doi:10.1038/nchembio.2238. (Epub ahead of print)

Kang Y, Tiziani S, Park G, Kaul M, Paternostro G (2014) Cellular protection using Flt3 and PI3 Kα inhibitors demonstrates multiple mechanisms of oxidative glutamate toxicity. Nat Commun 5:3672

Korytowski W, Basova LV, Pilat A, Kernstock RM, Girotti AW (2011) Permeabilization of the mitochondrial outer membrane by Bax/truncated Bid (tBid) proteins as sensitized by cardiolipin hydroperoxide translocation: mechanistic implications for the intrinsic pathway of oxidative apoptosis. J Biol Chem 286:26334–26343

Kwon MY, Park E, Lee SJ, Chung SW (2015) Heme oxygenase-1 accelerates erastin-induced ferroptotic cell death. Oncotarget 6:24393–24403

Lachaier E, Louandre C, Godin C, Saidak Z, Baert M, Diouf M, Chauffert B, Galmiche A (2014) Sorafenib induces ferroptosis in human cancer cell lines originating from different solid tumors. Anticancer Res 34:6417–6422

Li XX, Tsoi B, Li YF, Kurihara H, He RR (2015) Cardiolipin and its different properties in mitophagy and apoptosis. J Histochem Cytochem 63:301–311

Linkermann A, Skouta R, Himmerkus N, Mulay SR, Dewitz C, De Zen F, Prokai A, Zuchtriegel G, Krombach F, Welz PS, Weinlich R, Vanden Berghe T, Vandenabeele P, Pasparakis M, Bleich M, Weinberg JM, Reichel CA, Bräsen JH, Kunzendorf U, Anders HJ, Stockwell BR, Green DR, Krautwald S (2014) Synchronized renal tubular cell death involves ferroptosis. Proc Natl Acad Sci U S A 111:16836–16841

Liu Y, Wang W, Li Y, Xiao Y, Cheng J, Jia J (2015) The 5-lipoxygenase inhibitor zileuton confers neuroprotection against glutamate oxidative damage by inhibiting ferroptosis. Biol Pharm Bull 38:1234–1239

Lorincz T, Jemnitz K, Kardon T, Mandl J, Szarka A (2015) Ferroptosis in involved in acetaminophen induced cell death. Pathol Oncol Res 21:1115–1121

Louandre C, Ezzoukhry Z, Godin C, Barbare JC, Mazière JC, Chauffert B, Galmiche A (2013) Iron-dependent cell death of hepatocellular carcinoma cells exposed to sorafenib. Int J Cancer 133:1732–1742

Maguire JJ, Tyurina YY, Mohammadyani D, Kapralov AA, Anthonymuthu TS, Qu F, Amoscato AA, Sparvero LJ, Tyurin VA, Planas-Iglesias J, He RR, Klein-Seetharaman J, Bayır H, Kagan VE (2017) Known unknowns of cardiolipin signaling: the best is yet to come. Biochim Biophys Acta 1862:8–24

Mancias JD, Wang X, Gygi SP, Harper JW, Kimmelman AC (2014) Quantitative proteomics identifies NCOA4 as the cargo receptor mediating ferritinophagy. Nature 509:105–109

Matsushita M, Freigang S, Schneider C, Conrad M, Bornkamm GW, Kopf M (2015) T cell lipid peroxidation induces ferroptosis and prevents immunity to infection. J Exp Med 212:555–568

Nakamura T, Imai H, Tsunashima N, Nakagawa Y (2003) Molecular cloning and functional expression of nucleolar phospholipid hydroperoxide glutathione peroxidase in mammalian cells. Biochem Biophys Res Commun 311:139–148

Nomura K, Imai H, Koumura T, Arai M, Nakagawa Y (1999) Mitochondrial phospholipid hydroperoxide glutathione peroxidase suppresses apoptosis mediated by a mitochondrial death pathway. J Biol Chem 274:29294–29302

Nomura K, Imai H, Koumura T, Kobayashi T, Nakagawa Y (2000) Mitochondrial phospholipid hydroperoxide glutathione peroxidase inhibits the release of cytochrome c from mitochondria by suppressing the peroxidation of cardiolipin in hypoglycaemia-induced apoptosis. Biochem J 351:183–193

Ou Y, Wang SJ, Li D, Chu B, Gu W (2016) Activation of SAT1 engages polyamine metabolism with p53-mediated ferroptotic responses. Proc Natl Acad Sci U S A 113:E6806–E6812

Roggia MF, Imai H, Shiraya T, Noda Y, Ueta T (2014) Protective role of glutathione peroxidase 4 in laser-induced choroidal neovascularization in mice. PLoS ONE 9:e98864

Saito Y, Shichiri M, Hamajima T, Ishida N, Mita Y, Nakao S, Hagihara Y, Yoshida Y, Takahashi K, Niki E, Noguchi N (2015) Enhancement of lipid peroxidation and its amelioration by vitamin E in a subject with mutations in the SBP2 gene. J Lipid Res 56:2172–2182

Sakai O, Uchida T, Roggia MF, Imai H, Ueta T, Amano S (2015a) Role of glutathione peroxidase 4 in glutamate-induced oxytosis in the retina. PLoS ONE 10:e0130467

Sakai O, Uchida T, Imai H, Ueta T, Amano S (2015b) Role of glutathione peroxidase 4 in conjunctival epithelial cells. Invest Ophthalmol Vis Sci 56:538–543

Sakai O, Uchida T, Imai H, Ueta T (2016a) Glutathione peroxidae 4 plays an important role in oxidative homeostasis and wound repair in corneal epithelial cells. FEBS Open Bio 6:1238–1247

Sakai O, Yasuzawa T, Sumikawa Y, Ueta T, Imai H, Sawabe A, Ueshima S (2016b) Role of GPx4 in human vascular endothelial cells, and the compensatory activity of brown rice on GPx4 ablation condition. Pathophysiology S0928–4680(16):30068–30072

Sakamoto H, Imai H, Nakagawa Y (2000) Involvement of phospholipid hydroperoxide glutathione peroxidase in the modulation of prostaglandin D2 synthesis. J Biol Chem 275:40028–40035

Sakamoto H, Tosaki T, Nakagawa Y (2002) Overexpression of phospholipid hydroperoxide glutathione peroxidase modulates acetyl-CoA, 1-O-alkyl-2-lyso-sn-glycero-3-phosphocholine acetyltransferase activity. J Biol Chem 277:50431–50438

Sato H, Tamba M, Ishii T, Bannai S (1999) Cloning and expression of a plasma membrane cystine/glutamate exchange transporter composed of two distinct proteins. J Biol Chem 274:11455–11458

Schneider M, Förster H, Boersma A, Seiler A, Wehnes H, Sinowatz F, Neumüller C, Deutsch MJ, Walch A, Hrabé de Angelis M, Wurst W, Ursini F, Roveri A, Maleszewski M, Maiorino M, Conrad M (2009) Mitochondrial glutathione peroxidase 4 disruption causes male infertility. FASEB J. 23:3233–3242

Seiler A, Schneider M, Förster H, Roth S, Wirth EK, Culmsee C, Plesnila N, Kremmer E, Rådmark O, Wurst W, Bornkamm GW, Schweizer U, Conrad M (2008) Glutathione peroxidase 4 senses and translates oxidative stress into 12/15-lipoxygenase dependent- and AIF-mediated cell death. Cell Metab 8:237–248

Sengupta A, Lichti UF, Carlson BA, Cataisson C, Ryscavage AO, Mikulec C, Conrad M, Fischer SM, Hatfield DL, Yuspa SH (2013) Targeted disruption of glutathione peroxidase 4 in mouse skin epithelial cells impairs postnatal hair follicle morphogenesis that is partially rescued through inhibition of COX-2. J Invest Dermatol 133:1731–1741

Shimada K, Skouta R, Kaplan A, Yang WS, Hayano M, Dixon SJ, Brown LM, Valenzuela CA, Wolpaw AJ, Stockwell BR (2016) Global survey of cell death mechanisms reveals metabolic regulation of ferroptosis. Nat Chem Biol 12:497–503

Shindou H, Shimizu T (2009) Acyl-CoA: lysophospholipid acyltransferases. J Biol Chem 284:1–5

Skouta R, Dixon SJ, Wang J, Dunn DE, Orman M, Shimada K, Rosenberg PA, Lo DC, Weinberg JM, Linkermann A, Stockwell BR (2014) Ferrostatins inhibit oxidative lipid damage and cell death in diverse disease models. J Am Chem Soc 136:4551–4556

Smith AC, Mears AJ, Bunker R, Ahmed A, MacKenzie M, Schwartzentruber JA, Beaulieu CL, Ferretti F, FORGE Canada Consortium, Majewski J, Bulman DE, Celik FC, Boycott KM, Graham GE (2014) Mutations in the enzyme glutathione peroxidase 4 cause Sedaghatian-type spondylometaphyseal dysplasia. J Med Genet 51:470–474

Sun X, Ou Z, Xie M, Kang R, Fan Y, Niu X, Wang H, Cao L, Tang D (2015) HSPB1 as a novel regulator of ferroptotic cancer cell death. Oncogene 34:5617–5625

Tan S, Schubert D, Maher P (2001) Oxytosis: a novel form of programmed cell death. Curr Top Med Chem 1:497–506

Tobaben S, Grohm J, Seiler A, Conrad M, Plesnila N, Culmsee C (2011) Bid-mediated mitochondrial damage is a key mechanism in glutamate-induced oxidative stress and AIF-dependent cell death in immortalized HT-22 hippocampal neurons. Cell Death Differ 18:282–292

Torii S, Shintoku R, Kubota C, Yaegashi M, Torii R, Sasaki M, Suzuki T, Mori M, Yoshimoto Y, Takeuchi T, Yamada K (2016) An essential role for functional lysosomes in ferroptosis of cancer cells. Biochem J 473:769–777

Uchida T, Sakai O, Imai H, Ueta T (2016) Role of glutathione peroxidase 4 in corneal endothelial cells. Curr Eye Res 15:1–6

Ueta T, Inoue T, Furukawa T, Tamaki Y, Nakagawa Y, Imai H, Yanagi Y (2012) Glutathione peroxidase 4 is required for maturation of photoreceptor cells. J Biol Chem 287:7675–7682

Wang SJ, Li D, Ou Y, Jiang L, Chen Y, Zhao Y, Gu W (2016) 2016 Acetylation is crucial for p53-mediated ferroptosis and tumor suppression. Cell Rep 17(2):366–373

Wortmann M, Schneider M, Pircher J, Hellfritsch J, Aichler M, Vegi N, Kölle P, Kuhlencordt P, Walch A, Pohl U, Bornkamm GW, Conrad M, Beck H (2013) Combined deficiency in glutathione peroxidase 4 and vitamin E causes multiorgan thrombus formation and early death in mice. Circ Res 113:408–417

Yagoda N, von Rechenberg M, Zaganjor E, Bauer AJ, Yang WS, Fridman DJ, Wolpaw AJ, Smukste I, Peltier JM, Boniface JJ, Smith R, Lessnick SL, Sahasrabudhe S, Stockwell BR (2007) RAS-RAF-MEK-dependent oxidative cell death involving voltage-dependent anion channels. Nature 447:864–868

Yamaguchi H, Hsu JL, Chen CT, Wang YN, Hsu MC, Chang SS, Du Y, Ko HW, Herbst R, Hung MC (2013) Caspase-independent cell death is involved in the negative effect of EGF receptor inhibitors on cisplatin in non-small cell lung cancer cells. Clin Cancer Res 19:845–854

Yan S, Yang XF, Liu HL, Fu N, Ouyang Y, Qing K (2015) Long-chain acyl-CoA synthetase in fatty acid metabolism involved in liver and other diseases: an update. World J Gastroenterol 21:3492–3498

Yang WS, Stockwell BR (2008) Synthetic lethal screening identifies compounds activating iron-dependent, nonapoptotic cell death in oncogenic-RAS-harboring cancer cells. Chem Biol 15:234–245

Yang WS, SriRamaratnam R, Welsch ME, Shimada K, Skouta R, Viswanathan VS, Cheah JH, Clemons PA, Shamji AF, Clish CB, Brown LM, Girotti AW, Cornish VW, Schreiber SL, Stockwell BR (2014) Regulation of ferroptotic cancer cell death by GPX4. Cell 156:317–331

Yang WS, Kim KJ, Gaschler MM, Patel M, Shchepinov MS, Stockwell BR (2016) Peroxidation of polyunsaturated fatty acids by lipoxygenases drives ferroptosis. Proc Natl Acad Sci U S A 113: E4966–E4975

Yu Y, Xie Y, Cao L, Yang L, Yang M, Lotze MT, Zeh HJ, Kang R, Tang D (2015) The ferroptosis inducer erastin enhances sensitivity of acute myeloid leukemia cells to chemotherapeutic agents. Mol Cell Oncol 2:e1054549

Yuan H, Li X, Zhang X, Kang R, Tang D (2016) Identification of ACSL4 as a biomarker and contributor of ferroptosis. Biochem Biophys Res Commun 478:1338–1343

Immune Regulation by Dead Cell Clearance

Masato Tanaka and Gen Nishitai

Abstract When cell death occurs in vivo, cell corpses are not left untreated, but are recognized and engulfed by phagocytes, such as macrophages and dendritic cells. In the past, cell death had been considered the final process of a cell's life, and cell corpses had been viewed as debris that is simply to be cleared by phagocytes. Recently, however, it has become clearer that various biological responses are induced with dead cells as the starting point. Most of these biological responses followed by cell death are thought to be mediated by macrophages and dendritic cells. In this review, we present the overview of molecular mechanisms and biological significance of dead cell clearance.

Abbreviations

DCs	Dendritic cells
PS	Phosphatidylserine
MFG-E8	Milk fat globule-EGF 8
Tim-3	T-cell immunoglobulin mucin-3
Tim-4	T-cell immunoglobulin mucin-4
Gas6	Growth arrest-specific gene 6
TUNEL	TdT-mediated dUTP-biotin nick end labeling
HMGB1	High mobility group box protein 1

Contents

M. Tanaka (✉) · G. Nishitai
Laboratory of Immune Regulation, School of Life Science, Tokyo University of Pharmacy and Life Sciences, 1432-1 Horinouchi, Hachioji, Tokyo 192-0392, Japan
e-mail: mtanaka@toyaku.ac.jp

Current Topics in Microbiology and Immunology (2017) 403:171–184
DOI 10.1007/82_2015_472

Published Online: 13 August 2015

1 Introduction

In the human body, there are 60 trillion cells working in a cooperative manner to sustain life. Cells that have fulfilled their roles in each tissue, or such abnormal cells as cancer cells and virus-infected cells, are swiftly cleared via cell death. The removal of unnecessary or harmful cells via cell death is believed to play a vital role in the maintenance of homeostasis in living organisms. When cell death occurs in vivo regardless of the setting (i.e., whether physiological or pathological), cell corpses are not left untreated, but are recognized and engulfed by phagocytes, such as macrophages and dendritic cells (DC) (Henson et al. 2001; Lauber et al. 2003, 2004; Poon et al. 2014; Ravichandran and Lorenz 2007). In physiological conditions, cell corpses are engulfed by phagocytes so swiftly that they are hardly detected outside the phagocytes (McIlroy et al. 2000). Even in the pathological conditions, massive cell death is often followed by rapid clearance of cell corpses and tissue repair within a short period. For example, injection of dexamethasone results in massive cell death of immature T cells in mouse thymus. Soon after such cell death occurs in thymus, the corpses are rapidly cleared by thymic macrophages, and cellular composition of thymus recovers within 24 h (Scott et al. 2001). Ischemia–reperfusion injury causes necrotic cell death of epithelial cells in cortico-medullary border of kidneys, but swift clearance of injured cells results in tissue regeneration and recovery of kidney functions (Bonventre and Yang 2011). These observations prompt us to consider that rapid clearance of cell corpses is the essential first step for regeneration of injured tissues.

In the past, the engulfment of dead cells by phagocytes was thought to play a role merely in terms of corpse clearance. Recently, however, it has been revealed that macrophages that engulf dead cells, depending on the situation, can elicit a variety of biological responses. In particular, accumulated findings point to the important roles played by macrophages in the phagocytosis of dead cells, including the repair and regeneration of damaged tissue.

In this review, we outline the mechanism of phagocytosis of dead cells by macrophages and discuss the kinds of roles this mechanism plays in biological response following cell death, such as immune responses, inflammation, repair, and regeneration.

2 Phagocytosis of Dead Cells by Macrophages

When cell death occurs in a living organism, the corpses are quickly recognized and engulfed by phagocytes, such as macrophages, rather than left alone. It is most likely that the mechanisms of corpse clearance by phagocytes depend on the mode of cell death. But with regard to the mechanism of corpse clearance following the occurrence of cell death, only the analysis of apoptotic cases has seen progress. The molecular mechanisms by which phagocytes recognize and engulf apoptotic cells have been studied intensively since the late 1990s. Previous studies have found that phagocytes recognize and engulf phospholipids called phosphatidylserine (PS), which are exposed on the surface of apoptotic cells (Nagata et al. 2010). As PS serves as a marker when dead cells are subjected to phagocytosis by phagocytes, they are referred to as the "eat-me" signals. In living cells, PS is localized on the inner side of the cell membrane; however, when cells undergo apoptosis, PS is exposed to the extracellular face. Annexin V is well known to have an ability to bind PS specifically and frequently used to detect surface exposure of PS in apoptotic cell corpses by flow cytometry analysis. Early phase of apoptotic cell corpses exhibits Annexin V positive and PI negative, indicating the PS exposure to cell surface without increase in cell membrane permeability.

It was only recently that the underlying molecular mechanism has also become clearer. It has been assumed that the asymmetrical distribution of PS in the cell membrane of living cells involves flippase, which functions to help PS move from the exoplasmic face to the cytoplasmic face of the cell membrane, although the actual molecular state has long been unknown. Recent studies have reported that ATP11c and CDC50A play a substantial role in the asymmetric localization of PS (Segawa et al. 2014). Of these, ATP11c has been found to be cleaved by a caspase during apoptosis. This cleavage is thought to render it inactive as a functional flippase. In addition to the inactivation of flippase activity, it has also been reported that when cells undergo apoptosis, Xkr8 is activated by caspases during apoptosis and plays a critical role in active transportation of PS from the inner surface of the membrane to the outer surface (Suzuki et al. 2013). It is now understood that the inactivation and activation of these two enzymes cooperatively facilitate the exposure of PS on the outer surface of the membrane, allowing for recognition by macrophages.

While PS is the unique molecule as the "eat-me" signals exposed on apoptotic cell corpses, a large number of molecules have been reported as PS-binding molecules expressed by phagocytes. Some of these molecules, such as MFG-E8, Mer, and T-cell immunoglobulin mucin-3 and -4 (Tim-3 and Tim-4) are confirmed to be involved in apoptotic cell clearance in vivo (Fig. 1). Although differences in the roles of these molecules have yet to be clarified in detail, it has been reported that different molecules are used, depending on the types of macrophages in the organism. For instance, macrophages resident in the abdominal cavity engulf apoptotic cells via Mer and Tim-4. Mer has the ability to bind growth arrest-specific gene 6 (Gas6) and protein S, both of which exhibit binding activity of PS on the

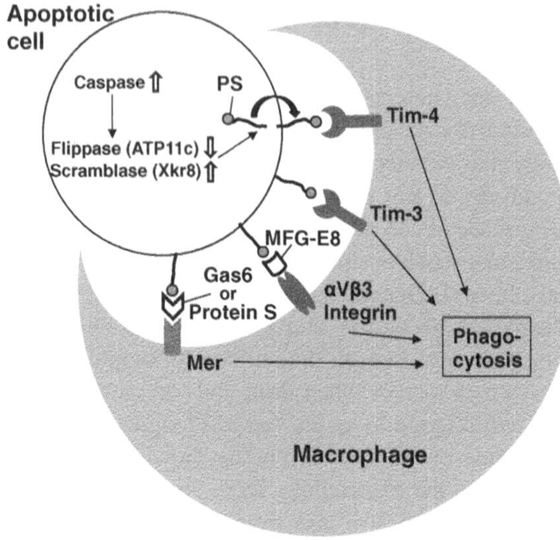

Fig. 1 Molecular mechanisms of apoptotic cell clearance by phagocyte

surface of apoptotic cells (Dransfield et al. 2015; Ishimoto et al. 2000; Nagata et al. 1996; Nakano et al. 1997; Zagorska et al. 2014), whereas Tim-4 can directly bind PS (Miyanishi et al. 2007). These two molecules coordinately play critical roles in efficient engulfment of apoptotic cells in peritoneal resident macrophages (Nishi et al. 2014). On the other hand, inflammatory macrophages induced by the intraperitoneal administration of thioglycollate engulf apoptotic cells in an MFG-E8-dependent manner (Hanayama et al. 2002). Tim-3 is expressed in splenic DCs, and the anti-Tim-3 antibody inhibits phagocytosis of apoptotic cells by CD8 + DCs and subsequently results in a reduced cross-presentation of apoptotic cell-associated antigens (Nakayama et al. 2009).

Consistent with the different expression of these molecules on phagocytes, gene-targeting mice of each molecule exhibit distinct phenotype. MFG-E8 is found mainly in the germinal centers of the spleen and lymph nodes and expressed in tingible body macrophages that engulf lymphocytes undergoing cell death. In MFG-E8-deficient mice, the abnormal phagocytosis of dead cells by these macrophages has been reported (Hanayama et al. 2004). On the other hand, in Mer-deficient mice, the clearance of apoptotic cell in thymus is impaired (Scott et al. 2001).

For cells that have undergone apoptosis to be swiftly engulfed by macrophages, it is imperative that macrophages migrate to the side of apoptotic cells. It has been reported that apoptotic cells release chemo-attractants called "find-me" signals to draw macrophages close. So far, there are reports suggesting that lysophosphatidylcholine and ATP released by apoptotic cells play an important role in accumulating macrophages (Chekeni et al. 2010; Lauber et al. 2003), but how these molecules actually function as find-me signals in vivo remains elusive.

Meanwhile, some studies have reported that molecules that are present on the surface of living cells inhibit phagocytosis by macrophages. These are referred to as the "don't-eat-me" signals, and CD47 molecules have been reported to possess this function (Oldenborg et al. 2000). However, whether CD47 on the surface of living cells actually inhibits phagocytosis is open to debate.

3 Apoptotic Cell Clearance in Living Organisms

In living organisms, the phagocytosis of apoptotic cells by macrophages occurs very quickly. Therefore, under physiological conditions, it is not usually possible to observe dead cells being left uncleared in any tissues. For instance, a substantial number of TUNEL-positive cells can be observed in thymus, but most of these cells are found to exist inside thymic macrophages in physiological conditions (McIlroy et al. 2000). So why is it that apoptotic cells must be cleared so quickly? Many studies have been carried out to address the significance of apoptotic cell engulfment in living organisms, mainly through the analysis of mice that lack molecules involved in the phagocytosis of dead cells by macrophages. In MFG-E8- and Mer-deficient mice mentioned above, the phagocytosis of apoptotic cells appears impaired, and serum anti-nuclear antibody and anti-DNA antibody titers show abnormal elevations (Hanayama et al. 2004; Scott et al. 2001). From these observations, the phagocytosis of apoptotic cells is believed to play a crucial role in the maintenance of immunological tolerance to self-antigen. To date, this phenomenon has been understood as such that phagocytes, by preventing autoantigens contained in dead cells from flowing out, inhibit the abnormal activation of autoimmune responses. However, another possibility has been pointed out that phagocytes might actually actively induce self-tolerance through the engulfment of dead cells. In other words, phagocytes that engulf dead cells are thought to transmit negative signals (deletion or anergy) to self-reactive T cells by presenting obtained autoantigens on MHC.

In multicellular organisms, a substantial number of tissue-resident cells, which contain tissue-specific self-antigens, undergo apoptosis constantly for turnover, and these apoptotic cells could become sources of tissue-specific self-antigens for antigen-presenting cells in each tissue. In fact, when cells undergoing apoptosis are intravenously administered, T-cell responses to antigens associated with dead cells are reportedly attenuated (Liu et al. 2002; Miyake et al. 2007a; Sun et al. 2004). The intravenously administered dead cells are phagocytosed by dendritic cells (DCs) in the spleen, and antigens associated with dead cells are presented to T cells for immunosuppression. Since the tolerance-inducing effects of apoptotic cells could be overcome when the DCs are stimulated by activation signals, the presentation of cell-associated antigens in the absence of costimulatory signals may lead to deletion or anergy of antigen-specific T cells (Liu et al. 2002).

On the other hand, under certain conditions, the phagocytosis of dead cells could result in the presentation of dead cell antigens by phagocytes, leading to the activation of T-cell responses to these antigens. Immune activation against dead cell-associated antigens has been extensively studied in the field of tumor immunity. It is reported that dead tumor cells, either killed in vivo or in injection of dead tumor cells, could activate tumor antigen-specific T-cell immunity under certain circumstances (Apetoh et al. 2007; Asano et al. 2011; Casares et al. 2005; Tesniere et al. 2008). The efficiency of tumor vaccination by dead tumor cells is largely depended on the nature of cell death in vaccinated tumor cells. It is also reported that calreticulin exposure on the dead tumor cells efficiently elicits anti-tumor immunity (Obeid et al. 2007). In another case, injection of artificial adjuvant vector cells expressing CD1d loaded with a-GalCer and tumor antigens elicits tumor immunity (Fujii et al. 2009; Shimizu et al. 2013). In this system, the vector cells are thought to undergo cell death in vivo and are phagocytosed by DCs. Then, DCs make a cross-presentation of tumor antigens to activate tumor antigen-specific CTLs in cooperation with activated NKT cells. Details of the mechanisms that define the direction of immune response to dead cell-associated antigens are still unknown, but if clarified, those mechanisms might shed light on new ways of controlling immune responses.

4 Subset of Macrophages Responsible for Phagocytosis of Apoptotic Cells

In living organisms, it is likely that macrophages and DCs control immune responses via the phagocytosis of apoptotic cells. So what kinds of cells are macrophages and DCs that actually play this role in living organisms? In each tissue in the living organism, there exist tissue-specific macrophages and DCs, and under physiological conditions, these phagocytes are thought to be responsible for the processing of dead cells. More recently, it has been revealed that these indigenous tissue-specific macrophages not only possess different properties depending on tissue, but also form several subpopulations in each tissue, with each playing a specific role (Gordon et al. 2014). This suggests the possibility that specific subpopulations are responsible for the phagocytosis of dead cells in tissues. Indeed, some subpopulations present in the spleen and lymph nodes have been reported to play a prominent role in the phagocytosis of dead cells, as well in associated immune responses. As described above, the intravenous injection of apoptotic cells induces immune tolerance to dead cell-associated antigens. In this case, marginal metallophilic macrophages and/or marginal zone macrophages are localized in the marginal zone of the spleen, i.e., the region where blood flows into the spleen, and (either one or both) have been shown to take up dead cells in blood (Miyake et al. 2007b). The critical role of these macrophages in the tolerance induction is proved by using the CD169-DTR mice, in which these macrophages can be specifically deleted by DT injection (Miyake et al. 2007b). CD11c-positive and CD103-positive dendritic cells are also localized in the marginal zone of the spleen, and they make cross-presentation of dead cell-associated antigens to CD8 T cells, demonstrating the coordinate immune regulation by macrophages and dendritic cells in the marginal zone (Qiu et al. 2009). Molecular mechanisms of tolerance induction by apoptotic cell infusion are also reported. Intravenous injection of apoptotic cells induces CCL22 expression in splenic metallophilic macrophages, resulting in the accumulation and activation of FoxP3 (+) Tregs (Ravishankar et al. 2014). On the other hand, as described above, massive cell death in tumor can induce immune activation to cell-associated antigens and activates anti-tumor immunity under certain circumstances. The candidate of macrophage/DC subset responsible for the immune activation associated with tumor cell death has been reported. When dead tumor cells are subcutaneously injected into mice, CD169-positive sinus macrophages localized in the lymphatic sinus of the lymph node take up dead cells or cell debris carried by the lymph flow and control immune responses to antigens associated with the dead cells (Asano et al. 2011). CD169-positive sinus macrophages consist of two subpopulations, CD11c-positive and CD11c-negative cells, and CD11c-positive cells, localized in the boundary border between sinus and T-cell zone, make a cross-presentation of dead cell-associated antigens to CD8 T cells. As immune responses to dead cell-associated antigens are closely related to the pathological conditions and the treatment of autoimmune diseases and cancer, the identification and functional analysis of involved macrophage subpopulations are essential research subjects.

5 Dead Cell-Derived Substances and Their Roles in Macrophage Activation and Regeneration

In the past, cell death had been considered the final process of a cell's life, and cell corpses had been viewed as debris that is simply to be cleared by phagocytes, such as macrophages, i.e., waste. Recently, however, as exemplified by the above-mentioned immune regulation by macrophages, it has become clearer that various biological responses are induced with dead cells as the starting point. Furthermore, we are beginning to understand how cells can actively regulate biological responses after cell death, by releasing physiologically active substances in the process of dying. Among the biological responses initiated by cell death, one of the most analyzed and advanced areas of research is inflammatory response. High mobility group box protein 1 (HMGB1) is one such endogenous stimulator of the immune system released from dead cells. HMGB1 was originally identified as a nuclear protein, but it is also passively released when cells undergo non-apoptotic cell death (Rovere-Querini et al. 2004; Scaffidi et al. 2002). When HMGB1 is once released, this is known to cause inflammation by acting on macrophages and DCs (Dumitriu et al. 2005; Messmer et al. 2004) (Fig. 2a). It is also reported that Mincle, a C-type lectin, is expressed by macrophages and recognizes SAP130 released from dead cells to induce sterile inflammation (Yamasaki et al. 2008). Such inflammation

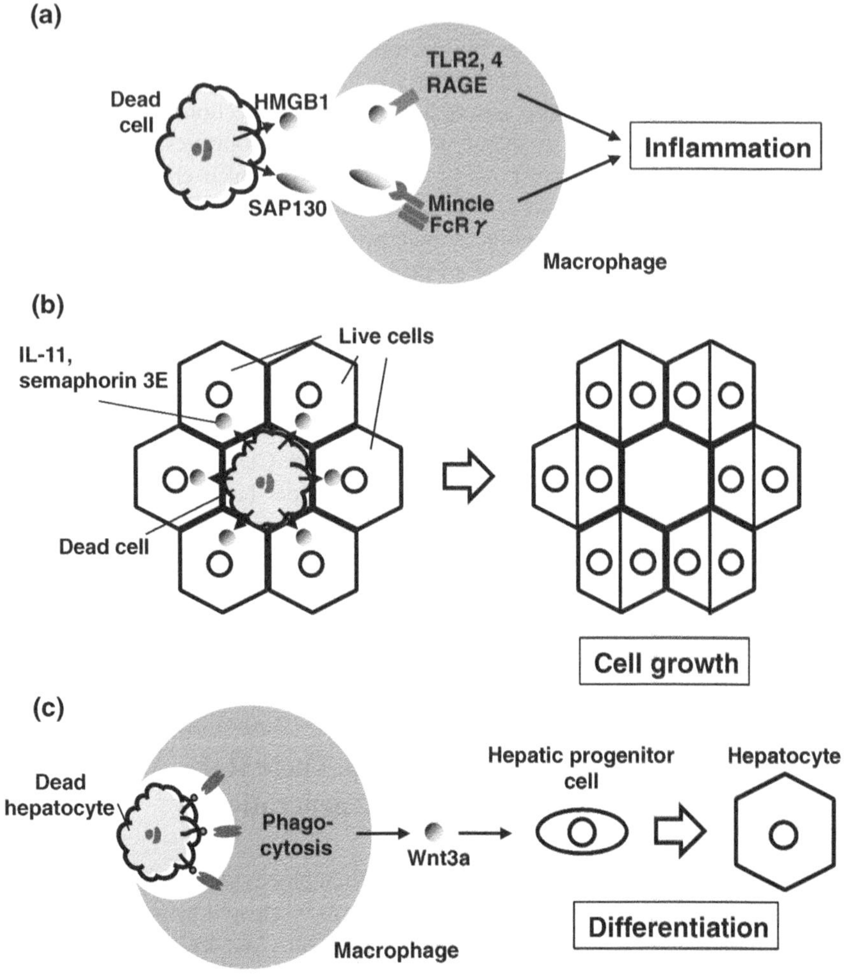

Fig. 2 Dead cell-derived substances and their roles in macrophage activation and regeneration

inducers that originate from dead cells are called damage-associated molecular patterns (DAMPs). Meanwhile, dead cell-derived substances have been reported to be involved in tissue repair and regeneration, in addition to inflammation. For example, IL-11, which is released by liver cells that had undergone cell death due to oxidative stress, has been reported to act on surrounding normal cells to promote liver regeneration (Nishina et al. 2012) (Fig. 2b). Similarly, some reports have indicated that semaphorin 3E is expressed in damaged liver cells and controls liver regeneration and fibrogenesis (Yagai et al. 2014). Furthermore, other reports have demonstrated that in chronic liver injury, phagocytosis of dead cells induces the expression of Wnt3a in macrophages, which contributes to regeneration by hepatic progenitors (Boulter et al. 2012) (Fig. 2c). Although cell death and regeneration seem strongly associated, the detailed mechanisms have not been clarified. Thus, in the future, progress of research in this area is highly anticipated.

6 Apoptosis and Non-apoptotic Cell Death in Vivo

In the past, apoptosis was thought to be the main mode of cell death occurring in vivo. Apoptosis is a type of cell death that occurs due to the activation of caspases in cells, resulting in the degradation of many intracellular substrates; it is an active death regulated by molecules. With regard to apoptosis, detection methods such as the TUNEL technique (Gavrieli et al. 1992) and activated caspase assays have already been established, and it is possible to detect apoptotic cells in situ. By using these methods, we clearly find that apoptotic cell death takes place in many organs during development and tissue turnover. Furthermore, a method to observe apoptosis in vivo in real time has also been developed, and the dynamics of apoptosis and its influence on surrounding cells are extensively studied during embryogenesis (Nonomura et al. 2013; Yamaguchi et al. 2011).

In contrast to apoptosis, non-apoptotic cell death, such as one that is caused by heat or other physical stimuli, or pathological cell death observed in various kind of diseases, used to be considered a passive form of cell death, and has been referred to as necrosis based on the morphological characteristics. Originally, it was believed that no special execution mechanisms existed in necrosis; however, in recent years, some modes of necrosis have been identified, which are controlled by molecular regulation. For instance, RIPK1/RIPK3- and MLKL-regulated cell death has been reported, which are referred to as necroptosis (Pasparakis and Vandenabeele 2015). It is also reported that caspase-1-regulated cell death is identified and referred to as pyroptosis (Lamkanfi and Dixit 2014). These non-apoptotic cell deaths exhibit morphological feature of classical necrosis, but especially in various pathological conditions, they appeared to contribute pathology of several diseases. More recently, another mode of cell death called ferroptosis, which requires iron ions, has been reported (Friedmann Angeli et al. 2014; Yang et al. 2014). The mechanisms of these various types of cell death have been clarified through analyses using cultured cells, and subsequent analyses of executing molecules in knockout mice have gradually unraveled their significance in vivo. Yet, when and how each type of cell death occurs in vivo has not been clarified. One of the reasons for the difficulties in analysis is the lack of methods to detect these new cell death events in vivo.

It is most likely that mechanisms of dead cell clearance by phagocytes depend on the mode of cell death. Furthermore, macrophages and dendritic cells could change the response to dead cells, depending on the mode of cell death. In order to explore the physiological and pathological consequence to cell death in vivo, we should carefully examine how macrophages and dendritic cells react to dead cells with different cell death modes.

7 Conclusion

In this review, we overviewed the mechanisms of dead cell clearance and its significance. Whereas studies to date have dramatically advanced the elucidation of molecular mechanisms apoptotic clearance, progress in research has generated new challenges, such as the identification of new cell death modes and the clarification of their significance and the determination of control mechanisms of biological response following cell death. The idea that cell death mechanisms simply exist to ensure cell removal might not fully explain the reason for the diversity in the modes of cell death. The hypothesis that each mode of cell death (purposely) elicits a specific biological response is attractive scientifically, but will require careful verification in the future through detailed analysis of cell death and subsequent biological response mechanisms.

Acknowledgments This work was supported in part by a Grant-in-Aid for Scientific Research (B) (26293089) from Japan Society for the Promotion of Science (JSPS), a Grant-in-Aid for Scientific Research on Innovative Areas (homeostatic regulation by various types of cell death) (26110006) from the Ministry of Education, Culture, Sports, Science and Technology (MEXT) in Japan, MEXT-Supported Program for the Strategic Research Foundation at Private Universities (2014–2019) in Japan, the Uehara Memorial Foundation, the Takeda Science Foundation, and the Naito Foundation. We thank T. Suito for secretarial assistance.

Competing interests: The authors declare that they have no competing interests.

References

Apetoh L, Ghiringhelli F, Tesniere A, Obeid M, Ortiz C, Criollo A, Mignot G, Maiuri MC, Ullrich E, Saulnier P, Yang H, Amigorena S, Ryffel B, Barrat FJ, Saftig P, Levi F, Lidereau R, Nogues C, Mira JP, Chompret A, Joulin V, Clavel-Chapelon F, Bourhis J, Andre F, Delaloge S, Tursz T, Kroemer G, Zitvogel L (2007) Toll-like receptor 4-dependent contribution of the immune system to anticancer chemotherapy and radiotherapy. Nat Med 13:1050–1059

Asano K, Nabeyama A, Miyake Y, Qiu CH, Kurita A, Tomura M, Kanagawa O, Fujii S, Tanaka M (2011) CD169-positive macrophages dominate antitumor immunity by crosspresenting dead cell-associated antigens. Immunity 34:85–95

Bonventre JV, Yang L (2011) Cellular pathophysiology of ischemic acute kidney injury. J Clin Investig 121:4210–4221

Boulter L, Govaere O, Bird TG, Radulescu S, Ramachandran P, Pellicoro A, Ridgway RA, Seo SS, Spee B, Van Rooijen N, Sansom OJ, Iredale JP, Lowell S, Roskams T, Forbes SJ (2012) Macrophage-derived Wnt opposes Notch signaling to specify hepatic progenitor cell fate in chronic liver disease. Nat Med 18:572–579

Casares N, Pequignot MO, Tesniere A, Ghiringhelli F, Roux S, Chaput N, Schmitt E, Hamai A, Hervas-Stubbs S, Obeid M, Coutant F, Metivier D, Pichard E, Aucouturier P, Pierron G, Garrido C, Zitvogel L, Kroemer G (2005) Caspase-dependent immunogenicity of doxorubicin-induced tumor cell death. J Exp Med 202:1691–1701

Chekeni FB, Elliott MR, Sandilos JK, Walk SF, Kinchen JM, Lazarowski ER, Armstrong AJ, Penuela S, Laird DW, Salvesen GS, Isakson BE, Bayliss DA, Ravichandran KS (2010)

Pannexin 1 channels mediate 'find-me' signal release and membrane permeability during apoptosis. Nature 467:863–867

Dransfield I, Zagorska A, Lew ED, Michail K, Lemke G (2015) Mer receptor tyrosine kinase mediates both tethering and phagocytosis of apoptotic cells. Cell Death Dis 6:e1646

Dumitriu IE, Baruah P, Valentinis B, Voll RE, Herrmann M, Nawroth PP, Arnold B, Bianchi ME, Manfredi AA, Rovere-Querini P (2005) Release of high mobility group box 1 by dendritic cells controls T cell activation via the receptor for advanced glycation end products. J Immunol 174:7506–7515

Friedmann Angeli JP, Schneider M, Proneth B, Tyurina YY, Tyurin VA, Hammond VJ, Herbach N, Aichler M, Walch A, Eggenhofer E, Basavarajappa D, Radmark O, Kobayashi S, Seibt T, Beck H, Neff F, Esposito I, Wanke R, Forster H, Yefremova O, Heinrichmeyer M, Bornkamm GW, Geissler EK, Thomas SB, Stockwell BR, O'Donnell VB, Kagan VE, Schick JA, Conrad M (2014) Inactivation of the ferroptosis regulator Gpx4 triggers acute renal failure in mice. Nat Cell Biol 16:1180–1191

Fujii S, Goto A, Shimizu K (2009) Antigen mRNA-transfected, allogeneic fibroblasts loaded with NKT-cell ligand confer antitumor immunity. Blood 113:4262–4272

Gavrieli Y, Sherman Y, Ben-Sasson SA (1992) Identification of programmed cell death in situ via specific labeling of nuclear DNA fragmentation. The Journal of cell biology 119:493–501

Gordon S, Pluddemann A, Martinez Estrada F (2014) Macrophage heterogeneity in tissues: phenotypic diversity and functions. Immunol Rev 262:36–55

Hanayama R, Tanaka M, Miwa K, Shinohara A, Iwamatsu A, Nagata S (2002) Identification of a factor that links apoptotic cells to phagocytes. Nature 417:182–187

Hanayama R, Tanaka M, Miyasaka K, Aozasa K, Koike M, Uchiyama Y, Nagata S (2004) Autoimmune disease and impaired uptake of apoptotic cells in MFG-E8-deficient mice. Science 304:1147–1150

Henson PM, Bratton DL, Fadok VA (2001) Apoptotic cell removal. Curr Biol 11:R795–R805

Ishimoto Y, Ohashi K, Mizuno K, Nakano T (2000) Promotion of the uptake of PS liposomes and apoptotic cells by a product of growth arrest-specific gene, gas6. J Biochem (Tokyo) 127:411–417

Lamkanfi M, Dixit VM (2014) Mechanisms and functions of inflammasomes. Cell 157:1013–1022

Lauber K, Blumenthal SG, Waibel M, Wesselborg S (2004) Clearance of apoptotic cells: getting rid of the corpses. Mol Cell 14:277–287

Lauber K, Bohn E, Krober SM, Xiao YJ, Blumenthal SG, Lindemann RK, Marini P, Wiedig C, Zobywalski A, Baksh S, Xu Y, Autenrieth IB, Schulze-Osthoff K, Belka C, Stuhler G, Wesselborg S (2003) Apoptotic cells induce migration of phagocytes via caspase-3-mediated release of a lipid attraction signal. Cell 113:717–730

Liu K, Iyoda T, Saternus M, Kimura Y, Inaba K, Steinman RM (2002) Immune tolerance after delivery of dying cells to dendritic cells in situ. J Exp Med 196:1091–1097

McIlroy D, Tanaka M, Sakahira H, Fukuyama H, Suzuki M, Yamamura K, Ohsawa Y, Uchiyama Y, Nagata S (2000) An auxiliary mode of apoptotic DNA fragmentation provided by phagocytes. Genes Dev 14:549–558

Messmer D, Yang H, Telusma G, Knoll F, Li J, Messmer B, Tracey KJ, Chiorazzi N (2004) High mobility group box protein 1: an endogenous signal for dendritic cell maturation and Th1 polarization. J Immunol 173:307–313

Miyake Y, Asano K, Kaise H, Uemura M, Nakayama M, Tanaka M (2007a) Critical role of macrophages in the marginal zone in the suppression of immune responses to apoptotic cell-associated antigens. J Clin Investig 117:2268–2278

Miyake Y, Asano K, Kaise H, Uemura M, Nakayama M, Tanaka M (2007b) Critical role of macrophages in the marginal zone in the suppression of immune responses to apoptotic cell-associated antigens. J Clin Investig 117:2268–2278

Miyanishi M, Tada K, Koike M, Uchiyama Y, Kitamura T, Nagata S (2007) Identification of Tim4 as a phosphatidylserine receptor. Nature 450:435–439

Nagata K, Ohashi K, Nakano T, Arita H, Zong C, Hanafusa H, Mizuno K (1996) Identification of the product of growth arrest-specific gene 6 as a common ligand for Axl, Sky, and Mer receptor tyrosine kinases. J Biol Chem 271:30022–30027

Nagata S, Hanayama R, Kawane K (2010) Autoimmunity and the clearance of dead cells. Cell 140:619–630

Nakano T, Ishimoto Y, Kishino J, Umeda M, Inoue K, Nagata K, Ohashi K, Mizuno K, Arita H (1997) Cell adhesion to phosphatidylserine mediated by a product of growth arrest-specific gene 6. J Biol Chem 272:29411–29414

Nakayama M, Akiba H, Takeda K, Kojima Y, Hashiguchi M, Azuma M, Yagita H, Okumura K (2009) Tim-3 mediates phagocytosis of apoptotic cells and cross-presentation. Blood 113:3821–3830

Nishi C, Toda S, Segawa K, Nagata S (2014) Tim4- and MerTK-mediated engulfment of apoptotic cells by mouse resident peritoneal macrophages. Mol Cell Biol 34:1512–1520

Nishina T, Komazawa-Sakon S, Yanaka S, Piao X, Zheng DM, Piao JH, Kojima Y, Yamashina S, Sano E, Putoczki T, Doi T, Ueno T, Ezaki J, Ushio H, Ernst M, Tsumoto K, Okumura K, Nakano H (2012) Interleukin-11 links oxidative stress and compensatory proliferation. Sci Signal 5:ra5

Nonomura K, Yamaguchi Y, Hamachi M, Koike M, Uchiyama Y, Nakazato K, Mochizuki A, Sakaue-Sawano A, Miyawaki A, Yoshida H, Kuida K, Miura M (2013) Local apoptosis modulates early mammalian brain development through the elimination of morphogen-producing cells. Dev Cell 27:621–634

Obeid M, Tesniere A, Ghiringhelli F, Fimia GM, Apetoh L, Perfettini JL, Castedo M, Mignot G, Panaretakis T, Casares N, Metivier D, Larochette N, van Endert P, Ciccosanti F, Piacentini M, Zitvogel L, Kroemer G (2007) Calreticulin exposure dictates the immunogenicity of cancer cell death. Nat Med 13:54–61

Oldenborg PA, Zheleznyak A, Fang YF, Lagenaur CF, Gresham HD, Lindberg FP (2000) Role of CD47 as a marker of self on red blood cells. Science 288:2051–2054

Pasparakis M, Vandenabeele P (2015) Necroptosis and its role in inflammation. Nature 517:311–320

Poon IK, Lucas CD, Rossi AG, Ravichandran KS (2014) Apoptotic cell clearance: basic biology and therapeutic potential. Nat Rev Immunol 14:166–180

Qiu CH, Miyake Y, Kaise H, Kitamura H, Ohara O, Tanaka M (2009) Novel subset of CD8 {alpha} + dendritic cells localized in the marginal zone is responsible for tolerance to cell-associated antigens. J Immunol 182:4127–4136

Ravichandran KS, Lorenz U (2007) Engulfment of apoptotic cells: signals for a good meal. Nat Rev Immunol 7:964–974

Ravishankar B, Shinde R, Liu H, Chaudhary K, Bradley J, Lemos HP, Chandler P, Tanaka M, Munn DH, Mellor AL, McGaha TL (2014) Marginal zone CD169 + macrophages coordinate apoptotic cell-driven cellular recruitment and tolerance. Proc Natl Acad Sci USA 111:4215–4220

Rovere-Querini P, Capobianco A, Scaffidi P, Valentinis B, Catalanotti F, Giazzon M, Dumitriu IE, Muller S, Iannacone M, Traversari C, Bianchi ME, Manfredi AA (2004) HMGB1 is an endogenous immune adjuvant released by necrotic cells. EMBO Rep 5:825–830

Scaffidi P, Misteli T, Bianchi ME (2002) Release of chromatin protein HMGB1 by necrotic cells triggers inflammation. Nature 418:191–195

Scott RS, McMahon EJ, Pop SM, Reap EA, Caricchio R, Cohen PL, Earp HS, Matsushima GK (2001) Phagocytosis and clearance of apoptotic cells is mediated by MER. Nature 411: 207–211

Segawa K, Kurata S, Yanagihashi Y, Brummelkamp TR, Matsuda F, Nagata S (2014) Caspase-mediated cleavage of phospholipid flippase for apoptotic phosphatidylserine exposure. Science 344:1164–1168

Shimizu K, Mizuno T, Shinga J, Asakura M, Kakimi K, Ishii Y, Masuda K, Maeda T, Sugahara H, Sato Y, Matsushita H, Nishida K, Hanada K, Dorrie J, Schaft N, Bickham K, Koike H, Ando T, Nagai R, Fujii S (2013) Vaccination with antigen-transfected, NKT cell ligand-loaded, human cells elicits robust in situ immune responses by dendritic cells. Cancer Res 73:62–73

Sun E, Gao Y, Chen J, Roberts AI, Wang X, Chen Z, Shi Y (2004) Allograft tolerance induced by donor apoptotic lymphocytes requires phagocytosis in the recipient. Cell Death Differ 11:1258–1264

Suzuki J, Denning DP, Imanishi E, Horvitz HR, Nagata S (2013) Xk-related protein 8 and CED-8 promote phosphatidylserine exposure in apoptotic cells. Science 341:403–406

Tesniere A, Panaretakis T, Kepp O, Apetoh L, Ghiringhelli F, Zitvogel L, Kroemer G (2008) Molecular characteristics of immunogenic cancer cell death. Cell Death Differ 15:3–12

Yagai T, Miyajima A, Tanaka M (2014) Semaphorin 3E secreted by damaged hepatocytes regulates the sinusoidal regeneration and liver fibrosis during liver regeneration. Am J Pathol 184:2250–2259

Yamaguchi Y, Shinotsuka N, Nonomura K, Takemoto K, Kuida K, Yosida H, Miura M (2011) Live imaging of apoptosis in a novel transgenic mouse highlights its role in neural tube closure. The Journal of cell biology 195:1047–1060

Yamasaki S, Ishikawa E, Sakuma M, Hara H, Ogata K, Saito T (2008) Mincle is an ITAM-coupled activating receptor that senses damaged cells. Nat Immunol 9:1179–1188

Yang WS, SriRamaratnam R, Welsch ME, Shimada K, Skouta R, Viswanathan VS, Cheah JH, Clemons PA, Shamji AF, Clish CB, Brown LM, Girotti AW, Cornish VW, Schreiber SL, Stockwell BR (2014) Regulation of ferroptotic cancer cell death by GPX4. Cell 156:317–331

Zagorska A, Traves PG, Lew ED, Dransfield I, Lemke G (2014) Diversification of TAM receptor tyrosine kinase function. Nat Immunol 15:920–928

The manufacturer's authorised representative in the EU is Springer Nature Customer Service Centre GmbH, Europaplatz 3, 69115 Heidelberg, Germany. If you have any concerns regarding our products, please contact ProductSafety@springernature.com

Printed and bound by CPI Group (UK) Ltd, Croydon, CR0 4YY
30/09/2026
02225195-0013